SILVER JEWELLERY MAKING

First published in 2021
Search Press Limited
Wellwood, North Farm Road
Tunbridge Wells, Kent TN2 3DR

Text copyright © Machi de Waard and
Janet Richardson, 2021
Photographs by Mark Davison at Search Press
Studios, except for: inside front cover photograph of
Machi de Waard and page 7, top right and bottom
left by Loren Stone (www.stonephotos.co.uk);
page 7 top left and bottom right by
Machi de Waard; reference photographs on
page 178 by Janet Richardson.

Photographs and design © Search Press Ltd, 2021

ISBN: 978-1-78221-735-0
ebook ISBN: 978-1-78126-667-0

The Publishers and authors can accept no
responsibility for any consequences arising from
the information, advice or instructions given in
this publication.

For details of suppliers, please visit the
Search Press website: www.searchpress.com

Visit the authors' websites at:
www.machidewaard.co.uk
www.janetrichardson.co.uk

ACKNOWLEDGEMENTS

Thanks to Beth Harwood, Mark Davison and all
staff at Search Press whose guidance, skill and
patience has shown no limits.

Machi:

I would like to thank: my parents, Marieke and
Frans, for everything; Dougall for being the peanut
gallery; my jewellery buddies, Cathy, Val and Sally
for inspiration and knowledge; Steve Richardson for
help with many things; my students for joining in
my love of jewellery and always having challenging
questions; and my friends who have supported me
over the years: Nick, Chris and Amanda, Martin
and Ange, Jog and Nong (best friend), Mike and
Jacqueline, Bob and Jane, Noel and Sarah, Jurgi
and Emma, Karen, Romy, Berend and Bea.

   Thank you to Janet for so many things that I can't
even list them and without whom this book would
not be possible.

Janet:

This book would never have existed had it not been
for the valued friendship and partnership of Machi.
Without the support of my husband, Steve, and my
family, I would not be the jeweller I am today. In this
wonderful world of jewellery the page is not long
enough to thank the inspiring people, students and
friends I have met along the way. I cannot thank you
enough for helping me get to this enjoyable place.

HEALTH AND SAFETY NOTICE

The techniques and skills demonstrated in this
book are intended to be undertaken with due
caution and diligence, and safety guidance is
provided throughout the book. The Publishers and
authors can accept no legal responsibility for any
consequences arising from the misapplication
of information, advice or instructions given in
this publication.

SILVER JEWELLERY MAKING

Machi de Waard & Janet Richardson

SEARCH PRESS

CONTENTS

 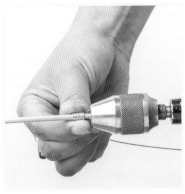 

INTRODUCTION 6

TOOLS AND MATERIALS 8
TOOLS **10**
MATERIALS **24**

SETTING UP 28
SETTING UP YOUR WORK AREA **28**
SETTING UP A BENCHPEG **30**
SETTING UP A VICE **31**

CORE SKILLS 32
USING A PIERCING SAW **34**
FILING **39**
SOLDERING **41**
PICKLING **52**
FINISHING **54**

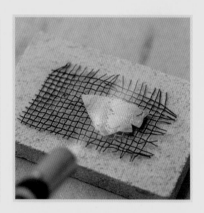

THE PROJECTS   58

1   **TWO SIMPLE RINGS**   **60**

2   **TWO TWIST RINGS**   **76**

    Additional Skills: Making Jump Rings   88

3   **STONE-CHIP EARRINGS**   **92**

    Additional Skills: Texturing Metal   104

4   **TEXTURED PENDANT**   **112**

    Additional Skills: Making a Bezel   122

5   **STONE SETTING FOR A RING**   **126**

6   **BROOCH WITH CUT CARD TECHNIQUE**   **146**

7   **FUSED PENDANT AND STUD EARRINGS**   **162**

MOVING FORWARDS   176

THE FOUR-STEP PLAN   **178**

GLOSSARY   184

APPENDICES   186

INDEX   192

INTRODUCTION

Silver Jewellery Making leads you through a series of seven tried-and-tested, structured projects. Each project introduces skills and techniques that will enable you to create a beautiful, finished piece of silver jewellery.

Between us, we have over fifty years of experience in jewellery making and teaching, so we know that understanding these skills will help you to move forwards with your own designs – whether you are already taking a class, coming to jewellery making anew, or looking to develop your techniques.

Our aim has not been to cover every possible technique in jewellery making. Instead, we focus on the key skills, going in depth to help you understand them better and enjoy them more. We also show you several ways of doing the same thing, such as ring sizing. The right method is the one that works for you.

A PRACTICAL APPROACH

Before we introduce the projects, we advise you on buying the tools and materials you will need, and provide information on setting up your own work area.

A Core Skills chapter, before the projects, is provided separately for easy reference. The projects themselves are designed to be made in succession, each one building on and adding to the skills you have previously learned.

- New, or additional, skills are explained in depth as they are introduced throughout the book. They are ideal as a point of reference to return to once you have completed the projects and begun to create your own designs.

- Handy Hints highlight the tricks and tips we have learned over the years and are keen to share with you.

- Troubleshooting solutions are included at the end of each project so that you know what to do next time if things have not gone to plan.

At the end of each project, we feature suggestions of designs for you to make as you progress, using the techniques you have practised. The last project includes pieces that you can make from your silver scrap – there will be scrap, however carefully you plan and buy!

The last chapter, Moving Forwards, is intended to inspire you to start designing your own pieces.

MAKING A LASTING IMPACT

Wherever you are on your journey into silver jewellery making, this book should become a resource to which you can keep returning for guidance and advice.

We hope that the fundamental skills and inspiration that this book will give you will lead you to enjoy jewellery making as much as we do.

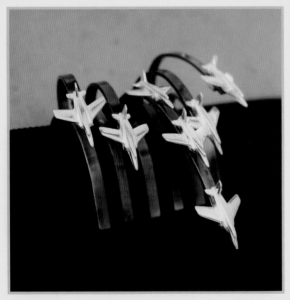

Top left

Sterling silver brooch with 18-carat yellow gold ball detail, by Machi de Waard.

Bottom left

Titanium and sterling silver brooch depicting the Red Arrow Gnats, by Janet Richardson.

Top right

Sterling silver concentric pendant with oxidized silver and gold plating, by Machi de Waard.

Bottom right

Carved sterling silver dragonfly neckpiece with sky-blue topaz, by Janet Richardson.

TOOLS AND

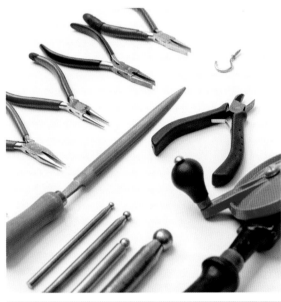

1   BENCH, HAND AND FORMING TOOLS_10

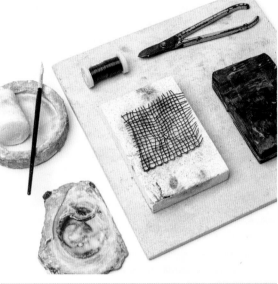

2   SOLDERING TOOLS_16

3   FINISHING AND POLISHING TOOLS_20

4   MISCELLANEOUS ITEMS_23

MATERIALS

TOOLS

On the following pages are our guidelines for buying and choosing your tools. We advise that you compare different suppliers and see what they have to offer. Be aware that it is often more economical to buy tools in sets or multiples. You can also use these guidelines to help you choose second-hand tools.

The tools listed are those you will need for the projects in this book. The optional tools on page 15 are useful but not essential.

BENCH, HAND AND FORMING TOOLS

ESSENTIAL BENCH, HAND AND FORMING TOOLS

1 Benchpeg and anvil (steel block)   Most of a jeweller's work is done on the benchpeg (**a**), which supports the jewellery while you are working on it. It also prevents damage to your fingers by creating a barrier between the tool and you. Benchpegs can be bought with a steel anvil (**b**), which is also used to attach it to a table. If you buy one without an anvil, it will need to be clamped down, for example with a G-clamp, shown on page 30 (it may be included). If you get one with a steel anvil, you do not need to get a separate steel block.

Benchpegs are often sold without the 'V' cut out of the centre – it can be cut out with a piercing saw (see page 30). The pegs can be used either way up – with the sloping side or the flat side uppermost. Traditionally, the sloping side is used for filing and the flat side is used for sawing.

The benchpeg and anvil are required for every project in this book.

2 Vice   A general-purpose, fixed, 75mm (3in) clamp-on table vice will suffice. Try to get one without teeth in the jaws, as it will mark the silver less when you clamp it.

3 Round ring mandrel   This is a heavy, tapered steel tool used for the forming of rings. Some mandrels have ring sizes marked on them and some are smooth. Those that have the sizes marked on can also double up as a ring stick (a ring sizing tool). However, a ring stick cannot double up as a mandrel because it is lightweight (usually aluminium), so it cannot be hammered against.

4 Hide mallet   A mallet is used to hammer metal into shape without marking it. These mallets are made from rolled hide with a varnish coating. When they are new, the varnish also covers the mallet face, so you will need to get rid of the varnish by hitting the mallet face on a rough surface – such as concrete or brick – and then filing it. If nothing happens, soak the mallet head in water to soften the varnish first.

There are also nylon and rubber mallets on the market, but we recommend the hide mallet, approximately 3.5cm (1⅜in) in diameter.

5 Ball pein (peen) steel hammer   It is fine to buy cheap hammers for use in silver jewellery making: you may even have these lying around. If you are buying a ball pein hammer second-hand, check that the ball has no flat faces on the round head and that the flat surface is not pitted.

6 Cross pein (peen) steel hammer   The narrower the cross pein, the finer the lines it will make. Again, if you are buying the hammer second-hand, check that the faces are not damaged or pitted.

7 Scribers   These are sharp, pointed steel tools used to mark your design or measurements on metal. The scriber is effectively your metalwork pencil: it makes accurate marks that cannot be rubbed out with your fingers.

8 Ruler   A ruler is an essential measuring tool. A steel ruler is best, as it is accurate.

GLOSSARY

Terms in **bold type** are explained in more depth later in the book, and in the glossary on pages 184–185.

5

6

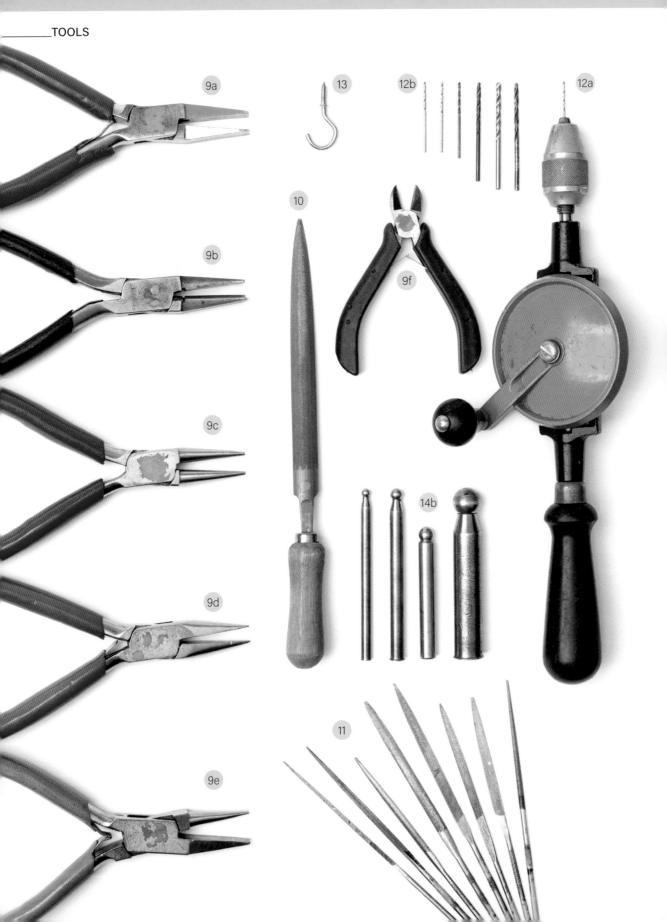

9a

13

12b

12a

9b

10

9f

9c

9d

14b

9e

11

9 Jewellery pliers These are used for the gripping and manipulation of metal. As with all pliers, check that the jaws come together along their whole length. You can buy sets of pliers, which usually include wire cutters, and these can be good value. It is better to have sprung pliers – this means that the handles open when you release your hand pressure. Jewellery pliers do not have teeth, as these would mark the metal.

The pliers listed below are the first ones we advise you to acquire, as they will be necessary for many of the projects. The distinctive ends of the pliers are illustrated below:

 9a) Flat-nose pliers;

 9b) Half-round/ring pliers;

 9c) Round-nose pliers;

9d) Chain-nose pliers (also known as snipe-nose pliers);

9e) Round-and-flat-nose pliers (these are optional);

9f) Wire cutters.

10 Half-round hand file (cut 2) with wooden handle Hand files are used for shaping and smoothing the surfaces and edges of metal. They come in a variety of shapes to suit different jobs and in different levels of coarseness. The higher the number, the finer the file.

Start with one good-quality, cut 2, half-round file. When you add to your tools, get a half-round cut 1 file next.

11 Needle files Needle files are small files that also come in different shapes and levels of coarseness. They can also be used for the shaping and smoothing of metal surfaces and edges, if a hand file is too big. You can buy needle files in sets, which are usually cut 2. Start with a set of steel needle files – you can also buy diamond needle files (see pages 39–40) but we

NOTE

It is usually the case that, when describing the coarseness of tools, the higher the number, the finer the tool.

recommend these as progression tools, as they are not essential.

12a) Hand drill The chuck is the most important part of the hand drill. When you look at the jaws, there should be no gap at the end when they are completely closed, so they can grip the small drill bits used in jewellery.

12b) Drill bits We recommend that you use drill bits ranging from 1–1.6mm (³⁄₆₄–¹⁄₁₆in) and no bigger than 2.5mm (³⁄₃₂in) when using a hand drill – otherwise you will have trouble drilling the hole. If you are going to buy one drill bit only, a good universal size is 1.2mm (³⁄₆₄in).

13 Cup hook This hook is perfect when twisting wire for rings (see page 79).

14a) Doming block (also called a dapping block) A doming block is used with doming punches (**14b**) to put regular curves into the surface of flat metal. Doming blocks are made in either nylon, wood, brass or steel. The steel ones are more robust. Some have shallow cups; some cups go as deep as half a sphere. You usually only buy this tool once, so a metal block with deep cups is a good purchase.

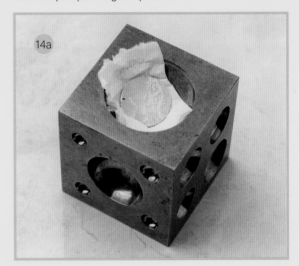

14b) Doming punches (also called dapping punches) Doming punches are used to form the metal into a curve alongside a doming block (above); they can also be used to texture the surface of the metal (see page 106). Doming punches are made of nylon, wood and steel and are usually sold in sets that don't include the larger sizes. If you wish to buy one larger punch to add to your set, get a steel one if your budget allows, as it will last longer.

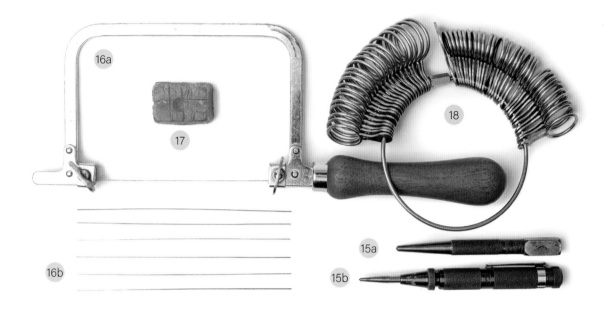

15 Centre punches   These are steel punches with a fairly blunt point. It is used to make a dimple in metal prior to drilling, and also to create a dotted texture (see page 106). We recommend a regular centre punch (**a**). (**b**) is an automatic centre punch that you do not need a hammer to use.

16a) Piercing saw or jeweller's saw   You can buy these saws with adjustable or fixed frames. The adjustable frames mean that you can use broken blades: simply adjust the frame to fit the length of the blade. The mechanism to tighten the nut onto the blade is either a screw fitting, a wing nut or a lever. It is easier to tighten a wing nut or lever than a screw. The saw that we use throughout the book has a fixed frame.

There are more expensive saw frames available which make tensioning the blade easier: these may be a consideration if you have issues with hand strength. Otherwise, we recommend a cheaper frame that has wing nuts or levers.

16b) Saw blades   These range from very coarse (8) to very fine (8/0). The average coarseness of blade for most jewellers is 2/0 or 3/0 (3/0 is finer) – see the chart on page 187 for more guidance. Saw blades are sold in bundles of twelve. They have a tendency to break, so do not worry if you break one. We recommend starting with 2/0 blades.

17 Beeswax or candle wax   This is used to lubricate the saw blade. Traditionally, beeswax is used, but candles do the job as well and you are more likely to have a candle to hand.

18 Ring sizers   These are available with or without half-ring sizes and are made of either metal or plastic. They also come suited to either wide or narrow rings. To fit the same size finger, a wider ring needs to be made bigger than a narrower ring, hence the two different types of ring sizers. There are also individual plastic ring sizers available that are adjustable. Any of these will do the job, but, if you buy a set, buy the narrow set first.

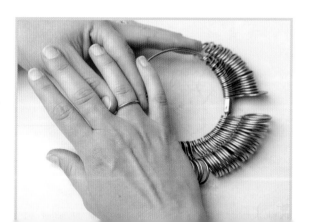

OPTIONAL BENCH, HAND AND FORMING TOOLS

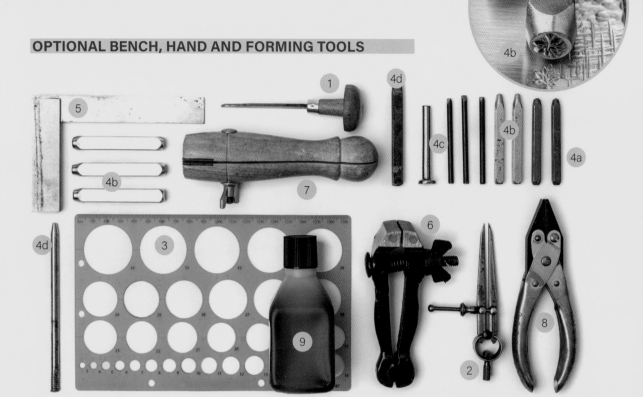

1 Flat scorper   This is a steel cutting tool with a flat, sharp end used for texturing, as shown on page 108.

2 Pair of dividers   Dividers are a steel measuring tool with sharp points, which is used to mark out lengths or circles. Invest in a good-quality pair so that they work accurately.

3 Circle template   Plastic shape templates – such as this circle template – are useful for transferring shapes onto metal with a scriber (see page 115).

4 Punches (also called metal stamps)   Punches are used to texture or pattern metal (see pages 105–107). There are many punches available commercially, such as number and letter punches (**4a**) and pattern punches (**4b**). You can also make your own pattern punch by cutting off the sharp end of a large round nail (**4c**) or steel rod or bar (**4d**).

5 Square (also called tri-square or engineer's square)   This is used to mark right angles and for checking the accuracy of a right angle.

6 Hand vice   This is a small, metal hand-held vice with a screw fitting at the side to tighten the jaws. It is used to grip metal securely.

7 Ring clamp   This wooden clamp holds items in place, usually rings, while you are working on them, for example to hold a ring securely while setting a stone. There are three main types of clamp: two hold the ring vertically (one with a screw to tighten, one with a wedge) and one holds it horizontally. We advise that you get the clamp that holds the ring vertically.

8 Parallel pliers   The jaws of these versatile pliers open in parallel – think of them as a small hand vice. They are strong and, as they hold the metal along the length of the jaws, they are less likely to mark the metal. These are a great addition to your tools as they hold small pieces of metal for filing and have strong jaws which are good for manipulating thick metal.

9 Platinol (oxidizing solution)   This is a chemical solution that blackens – or oxidizes – silver. Always use it with gloves and goggles (see page 29) and read the manufacturer's instructions.

SOLDERING TOOLS

1 Torches   A torch uses gas to produce a flame that is used to heat silver in order to solder together pieces of silver.

You have several options when it comes to choosing a torch: we suggest that you start with a hand-held torch because it is inexpensive. It is important that the torch has a big enough flame to heat the work. Smaller flames are suitable for **soldering** jump rings and rings, but will not cope with bangles or the cut card brooch project in this book (see pages 146–161).

The three hand-held torches featured have big enough flames to heat your work. There are other options, including a gas torch (**1c**), which is attached to a propane gas bottle with a hose. Torches use either butane or propane gas or a butane–propane mix. All gases are fine, but propane burns hotter than butane. See below for more details:

1a) Hand-held jeweller's torch/crème brûlée kitchen torch   This is a hand-held jeweller's torch with a suitably large flame. It is easily refilled with butane cigarette lighter gas (**1d**). If you choose this option, ensure that the description says that it can handle most soldering jobs. Avoid torches with 'mini' or 'micro' in the description as they will not get hot enough.

1b) Plumber's or DIY torch   This is also a hand-held torch. The torch is screwed onto the top of either a butane or butane–propane mix canister (**1e**). As the torch is screwed to the gas canister, these work until the gas runs out – the canister can be replaced with a new one. It is important to buy the right size torch as some create a flame that is too big.

When the canister is new, do not tip the canister at more than a forty-five-degree angle – the gas can run down the nozzle and create a big flame.

1c) Gas torch with propane gas bottle   This is an option to consider after you have had some experience and are making more jewellery. The torch is attached by a reinforced rubber hose to a propane gas bottle using a pressure regulator. You can buy different-size burners that result in different-size flames. If you buy a Sievert brand burner, we recommend sizes 3939 or 3940. Do not mix brands, as they may not be compatible.

For full details, follow the manufacturer's set-up guidelines and health and safety advice.

2 Lighter   The lighter is used to light the gas from the torch (see page 47). For the purposes of safety, make sure that the lighter is not near the flame while soldering.

The lighter shown is refillable with butane cigarette lighter gas and produces a flame. There are also piezoelectric lighters available, which produce a spark and do not need gas to work. Other options are cigarette lighters, battery-powered automatic torch lighters and flint torch strikers.

3 Steel tweezers and reverse-action tweezers
Steel tweezers (**3a**) are used to hold and manipulate your metals whilst soldering (see pages 41–51). Steel tweezers are ideal as solder will not flow onto them. Steel reverse-action tweezers (**3b**) are available with straight or curved ends: either is fine. They open when you squeeze them and spring closed.

4 Plastic, brass or copper tweezers (tongs)
These are used to remove your work from **pickle** (see pages 52–53).

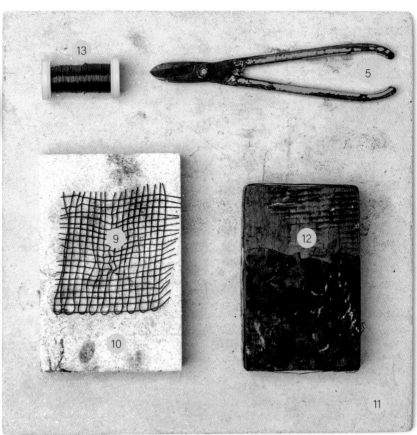

5 Tin snips (shears)　These are used to cut sheet metal and trim solder into **pallions**. We recommend the models where the handles do not come together at the back, as when the handles close they can pinch your skin and cause blood blisters.

As a general rule, it is best to use tin snips to cut sheet and use wire cutters or a saw to cut wire.

6 Borax　**Borax** is a type of **flux**. Flux prevents oxides forming on the metal during heating, keeping it clean. There are several brands and varieties of flux – we recommend starting with borax. It is inexpensive and available in either powder or solid cone form, as seen here. It is mixed with water to form a solution.

7 Borax dish and b) slate　If you are using a borax cone, you need to grind it with water in either a borax dish (**a**) or on a piece of slate (**b**). When you buy a new borax dish, it will take a bit of time to bed it in because it is porous. Also, because the dish is porous, you may wish to put it in a glass bowl or saucer.

8 Paintbrush　Use a brush to paint the borax onto silver. Any small brush is fine.

9 Steel mesh (optional)　A steel mesh is used on a soldering block to raise the work so that heat can go underneath it. It can come with two edges bent down as legs. Avoid very fine mesh, as it won't last.

10 Soldering block　This is a general-purpose soldering surface, made from asbestos substitute, that reflects the heat back onto your work. It is inexpensive and lasts a long time.

11 Large soldering sheet (fireproof sheet)
300 × 300mm (12 × 12in). This sheet goes under any soldering surface to increase the size of the fireproof surface area for safety.

12 Charcoal block (optional)　This is also used as a soldering block. There are natural and compressed charcoal blocks available, both of which have heat-reflective properties.

13 Binding wire　This is iron or steel wire which is used to bind items together for soldering. It can also be used for texturing (see pages 107 and 116). It comes in a variety of thicknesses. Remember to remove the wire before pickling your work, as steel should not go in the pickle.

14 Slow cooker (crock pot)　This is one way in which you can heat **pickle** (see below) – an inexpensive slow cooker is fine for this purpose. See page 52 for instructions.

15 Pickle　Pickle comes as a powder that is mixed with water and called safety pickle as it is much safer to use than the sulphuric acid used in the past. Safety pickle, which is sodium bisulphate, is used to clean oxides and flux off the metal after soldering. It needs to be heated to work effectively (see page 52). Sodium bisulphate can be disposed of safely. You can use any acidic liquid – such as cola or vinegar – to remove oxides, but they won't be as efficient in removing the flux residue.

16 Bowl of water　Water is used to rinse tweezers and silver jewellery work immediately after it has been removed from the pickle.

Grind the borax cone (6) with water in a borax dish (7a).

FINISHING AND POLISHING TOOLS

A brass brush is the cheapest polishing tool that can be used for polishing all the projects in this book. A few other options for polishing are listed here. See pages 54–57 for information to help you choose.

1 Burnisher   A burnisher is used to compress or move metal without marking it, when setting stones or highlighting areas of your work. Burnishers come in short or long lengths of steel and can be curved or straight. It doesn't matter if you get a straight or a curved one, but a shorter one might be easier to use initially.

You can also use the back of the bowl of a steel teaspoon as a burnisher (see page 57).

2 Brass brush   A brass brush is used to scratch-polish metal. Brass brushes have wooden handles and brass bristles. The bristles should not be too tough, so buy your brush from a jewellery supplier to ensure that you get softer bristles. Brass brushes come in various sizes, any of which are suitable.

3 Abrasive paper   Abrasive paper is used to sand out scratches and has a variety of names, including emery paper and wet-and-dry paper. These papers come in different grades from coarse to fine. Common grades for jewellery making are 240 (coarse), 500 or 600 (medium) and 1200 (fine). Start with these grades then try others as necessary. Wet-and-dry paper, which we use throughout this book, is so-called because it can be used dry or with water to aid abrasion.

4 Sanding (emery) sticks   Sanding sticks are used to remove marks to work towards a polish. These are available to buy pre-made in various levels and grades of coarseness, and are usually 30cm (12in) long – rectangular or round – and approximately 1cm (⅜in) thick. You can make your own sanding sticks using wooden sticks such as dowelling with abrasive paper wrapped around them. Instructions can be found on page 55.

5a) Barrel polisher   This is used to put a final, high polish on your pieces. There are different types of barrel polisher available: we use either an Evans rubber barrel polisher or a Pioneer mini octagonal barrel polisher, shown above, right. Barrel polishers are filled with steel shot (**b**) and water with a barrelling compound – a soap with rust inhibitor (such as Barrelbrite). The barrel is put on a motor which turns the barrel and the steel shot burnishes the silver, polishing it. Instructions are found on pages 54 and 56.

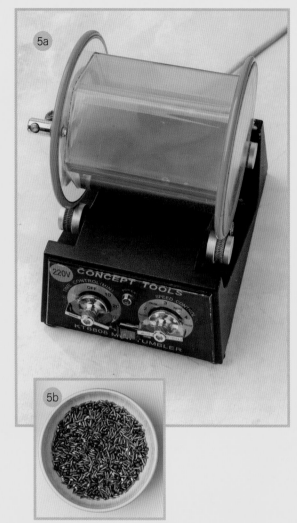

5b) Steel shot   Steel shot, which is used inside the barrel polisher (above), is either steel or stainless steel; the advantage of stainless steel is that it won't rust – keep the shot in the liquid in the barrel. Steel shot comes in different forms: as balls or in a mixture of shapes.

6a) and b) Sieve and jug   These are used together to empty the contents of the barrel polisher once it has finished. The sieve (**a**) separates the liquid from the shot and from your work; the jug (**b**) sits under the sieve to catch the liquid.

7 Bowl of water (not shown)   Once you have taken the work out of the barrel polisher, it will need to be rinsed in water. Dry your piece on kitchen paper after rinsing to avoid getting water marks on the metal.

10b

Rouge

10a Tripoli

8 Tripoli   This is a coarse brown polishing compound that comes in a block. Tripoli is used to remove sanding scratches and fine file marks. In the projects it is used in conjunction with a buff stick (**10a**).

9 Rouge   This is a fine red polishing compound that also comes in a block. It is usually used for a final polish after Tripoli has been used. Rouge is also used with a buff stick (**10b**).

10 Buff sticks (polishing sticks)   These are used for polishing with compounds such as Tripoli and rouge. These sticks are made with either felt or chamois leather. The chamois is used only with the finest abrasive polishes. The felt sticks can be used with both Tripoli (**8**) and rouge (**9**). Each polish needs to be applied to a separate stick – do not mix polishes on a single stick.

11 Abrasive rubber block   Abrasive blocks are available in different grades from fine to very coarse. These blocks have the same abrasion as abrasive paper but in solid rubber form – they can also be used for sanding. The very coarse block can also be used to give a satinized finish (see page 109).

12 Polishing cloths   These cloths are impregnated with polishing compound to give a final shine to a silver surface.

13 Brasso   This is wadding impregnated with polish that is rubbed onto the surface of the metal and is then rubbed off with a cloth or a sheet of kitchen paper. Brasso is slightly more abrasive than Silvo (which is specifically manufactured for silver), which makes it ideal for a final, fine sanding before you move on to use a polishing cloth.

MISCELLANEOUS ITEMS

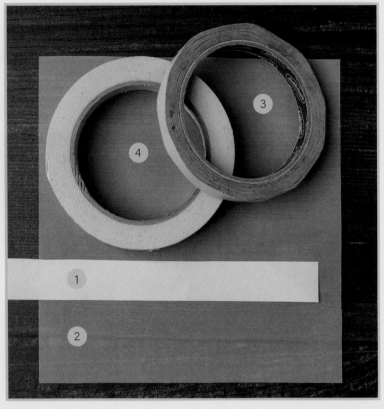

1 Strip of paper Two of these are used in project 1 (see pages 62 and 64) to help you work out the length of silver wire for your two simple rings.

2 Tracing paper Use tracing paper to trace your design and then transfer it to your silver sheet in project 6 – Brooch with cut card technique (see pages 148–149).

3 Double-sided tape Use to hold down a cabochon stone while working out the length of a bezel strip (see page 123), and along with the tracing paper to transfer your brooch design to silver sheet in project 5 (see page 148).

4 Masking tape This versatile tape is used in project 4 for holding down wire for texturing (see pages 107 and 116) and project 7 for attaching thread to the pendant to work out where to position the bail (see page 169); it is also used in the making of abrasive sanding sticks (see page 55).

MATERIALS

It is more economical to buy all the materials needed for the projects in this book at once than to buy them separately. Do compare different suppliers' prices as well. The materials you will need are shown over the following four pages.

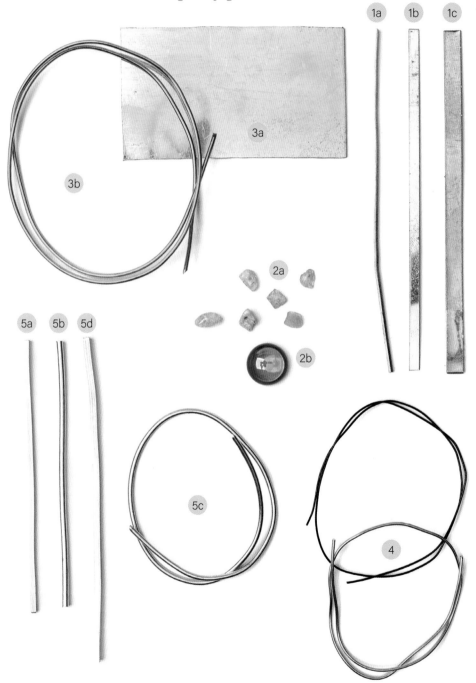

Specialist jewellery suppliers stock the materials needed to make jewellery. Unless you live near a supplier, you will have to order these online. Ordering the specific materials for these projects will give you an understanding of what you are buying and what the materials look like, to enable you to purchase more for your own work.

Cookson Gold (cooksongold.com) in the UK, Rio Grande (riogrande.com) in the USA and Karl Fischer (goldschmiedebedarf.de) in Germany are large, international jewellery suppliers, but there are many more, the details of which can be found on the Search Press website.

1 Silver solder There are three grades of silver **solder**: medium (**1a**), easy (**1b**) and hard (**1c**) solder – see page 41 for an explanation of these grades.

2 Stones To complete the projects in this book, you will need six semi-precious stone chips (**2a**) with holes drilled in them. These are often sold in strings rather than individually. You will also need a 10mm (⅜in) round cabochon cubic zirconia stone (**2b**) in a colour of your choice. (We got this stone from Ward Gemstones – www.wardgemstones.com)

3 Copper Copper sheet (**a**), 0.7mm (21-**gauge**) thick and at least 65 × 35mm (2½ × 1⅜in).

Copper wire (**b**), no thicker than 1mm (¹⁄₁₆in) no less than 30cm (12in) long, for practising wire wrap loops and jump rings.

4 Wire for texturing This can be binding wire (see page 19). It can also be copper, silver or brass wire approximately 0.8mm (20-gauge) thick or any piece of scrap wire you have lying around. A 16cm (6¼in) length is ample for the projects in this book.

5 Silver for rings These lengths are for average ring sizes. Add 10mm (⅜in) to the length for above-average ring sizes. The lengths listed will make rings of most sizes except the very largest – i.e. above UK size V, US size 10¾. Add 10mm (⅜in) to the length of the wire for above-average ring sizes.

5a) 75mm (3in) of 1.5mm (15-gauge) square sterling silver wire;

5b) 75mm (3in) of 2mm (12-gauge) round sterling silver wire;

5c) 210mm (8¼in) of 1.5mm (15-gauge) round wire;

5d) 90mm (3½in) of 2mm (12-gauge) square wire.

Below, a selection of stones that you might like to buy if you progress with stone settings – see pages 142–145.

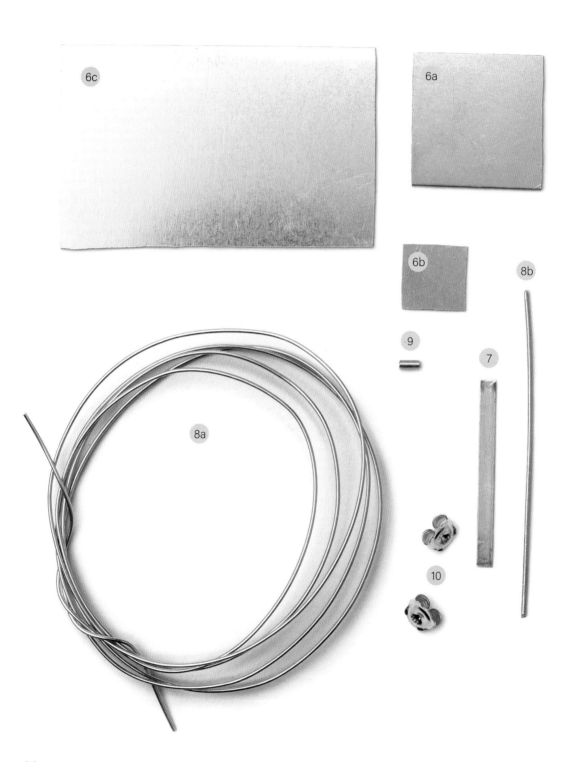

6 Silver sheet

6a) 25 × 25mm (1 × 1in) of 0.8mm (20-gauge) thick silver sheet;

6b) 12 × 12mm (½ × ½in) piece of 0.5mm (24-gauge) thick sterling silver sheet;

6c) 60 × 40mm (2⅜ × 1⁹⁄₁₆in) of 0.6mm (22-gauge) sheet.

A NOTE ABOUT SILVER SHEET

When you purchase silver sheet, it comes with a plastic coloured film over it which is there to protect the surface from scratches (in the photograph on page 9 it is blue). It peels off, but sometimes it is useful to leave it on to protect the surface.

However, make sure that the film is removed from the silver before you do any heating.

SILVER SCRAP

When you are making jewellery, you will inevitably have pieces of silver scrap and silver filings left over. It is important to keep these – not only because they have value, but also because it can be used to make more jewellery, as in project 7 (see pages 162–175).

Scrap can be sold back to bullion dealers once you have accumulated enough, but it is bought at a reduced price, so it is more fun and economical to use it in your work.

7 Fine silver bezel strip

35mm (1⅜in) of 3mm (9-gauge) wide × 0.3mm (28-gauge) thick **bezel** strip.

8 Silver wire for other components

8a) 1m (40in) of 0.8mm (20-gauge) round wire. This is more than you need for the projects; however, the excess will allow for any errors.

If the stone chips have especially small holes, you will also need 30cm (12in) of 0.6mm (22-gauge) round wire.

8b) 60mm (2⅜in) of 0.9mm (19-gauge) round silver wire.

9 Silver tube for the brooch catch (project 6)

5mm (³⁄₁₆in) of silver tube with an outside diameter of 1.6mm and an inside diameter of 1mm. If you cannot get this exact size, go up one size, not down in size.

10 Two medium-weight scroll backs
These are attached to the posts of the earrings in project 7 – Fused pendant and stud earrings (see page 171).

SETTING UP

Where you set up your workbench will be unique to you. These pages will give you guidelines for setting up your tools in a logical and safe way.

SETTING UP YOUR WORK AREA

You can set up in a small area to work as a jeweller. The important thing is that the table you are working on is stable, secure and not your best dining table! You should be able to clamp your benchpeg (1) to the edge of the table (see overleaf).

A lamp (2) to light up the area on the benchpeg will help you to see more clearly what you are doing. Consider the height at which you are sitting: you will sit lower at a jeweller's bench than you would at a normal table. It is important that you are comfortable as you work and not hunched over. An adjustable chair will help you to find a comfortable height. As a guideline, lower your chair so that your work surface comes to between your elbows and your shoulders.

It is advisable to keep all the soldering items on one side (3) and hand tools (4) on the other side of your work area. This protects the tools, as the fumes from the pickle (5) can cause some tools to rust. Place the soldering items on a fireproof surface, such as a soldering sheet (6).

The more you work at your bench, the more you will need to adjust it to suit you, but this is a good starting point.

OTHER POINTS TO CONSIDER

- It is best not to set up near a window, so that you can see the colour of the metal when you solder: you can see it better in a darker area. In bright light, it is hard to see the flame of the torch or the colour of the metal as it heats.

- Everything that you need for soldering should be in one area so you do not need to walk around with hot metal or metal with chemicals on it. The soldering area should also be clear of flammable materials.

- It is a good idea to put the water, flux and pickle pot on metal trays (baking trays are good for this purpose) to protect your table top and make cleaning up easier.

- Put a note somewhere you will see it every time you leave your work station, reminding you to turn off all the equipment. You do not want to leave your pickle on overnight!

HEALTH AND SAFETY

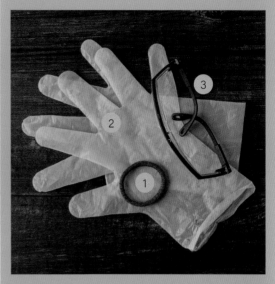

Before starting any of the projects, tie back long hair with a hair tie (1), remove any dangling jewellery or scarves and put on sensible shoes (i.e. not sandals, high heels or open-toed shoes) that will protect your feet if something hot or heavy falls on them.

Wherever a manufacturer's instructions indicate, you should wear protective equipment. For instance, always wear protective gloves (2) and goggles (3) when handling the oxidizing solution Platinol.

We advise that you wear an apron, as your clothes could get dirty. If you have a leather apron, it is a good idea to lay this across your lap if you are soldering while sitting down.

You can also wear goggles while you are soldering, but make sure that they are clear so you can see properly.

Relevant health and safety advice will be mentioned when necessary throughout the projects. Keep and read all manufacturers' instructions, and read the health and safety warnings on your tools and materials. All chemicals come with safety data sheets – alternatively, these can be requested from the manufacturer.

SETTING UP A BENCHPEG

Most of a jeweller's work gets done on the benchpeg. It supports the jewellery while you are working on it. It also prevents damage to your fingers by creating a barrier between the tool and you. It is best attached to a stable table, but be aware that the clamp may mark the bottom of the table.

NOTE

Over time, you may need to modify your benchpeg to suit the work you are making. Every benchpeg gains a unique character as it is used. This is Machi's benchpeg, below, after many years of use.

Sawing the benchpeg

Benchpegs are often sold without the 'V' cut out of the centre. If that is the case, cut it out with a piercing saw, as shown in the photographs above.

Attaching the benchpeg to the anvil

Benchpegs can be bought with a steel anvil, which is also used to attach it to a table. If you get one with a steel anvil, you do not need to get a separate steel block. The photograph above shows how the benchpeg is placed in the anvil.

Attaching the anvil

The anvil has a built-in clamp that you tighten to attach it to a table. The pegs can be used either way up – with the sloping side or the flat side uppermost. Traditionally, the sloping side is used for filing and the flat side is used for sawing. For general use, it is best to keep the benchpeg flat side up.

Other anvil or clamp options

A benchpeg without an anvil can be attached to a table with a G-clamp. Other types of benchpegs are available, often with cut-outs of various sizes and shapes already made. It is the cut-out 'V' that is the most important feature.

If your benchpeg does not have an anvil, you will need to get a separate steel block.

SETTING UP A VICE

A vice is a versatile tool with movable jaws that are used to hold tools or materials firmly in place while working. A vice is usually attached to a workbench with an integral clamp, or bolted down.

A good vice to start with is a general-purpose, fixed, 75mm (3in) clamp-on table vice.

This vice has a built-in clamp that you tighten to attach it to a table.

Here the vice is clamped to the side of a stable table.

MAKING SOFT JAWS

If the vice has teeth, it is useful to make soft covers for the jaws to avoid marking metals that are placed in the vice. Soft jaws can be made from copper or aluminium, as shown below. Another alternative is to use leather or a folded sheet of kitchen paper wrapped around a piece before it is clamped in the jaws.

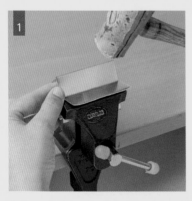

1. To make metal soft jaws, cut two rectangles of metal to the length of the vice jaws and 3cm (1³⁄₁₆in) wide. Put one piece in the vice lengthways so half of the rectangle sticks out above the jaw. Using a mallet, hammer it over the jaw to form it into an 'L' shape.

2. Repeat with the second piece. Now you have a pair of soft jaws for the vice.

Rubber ready-made soft jaws can be bought for some vices, as shown in this photograph.

CORE SKILLS

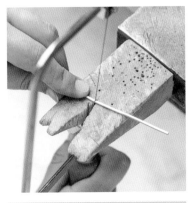

1   USING A PIERCING SAW_34

2   FILING_39

3   SOLDERING_41

4   PICKLING_52

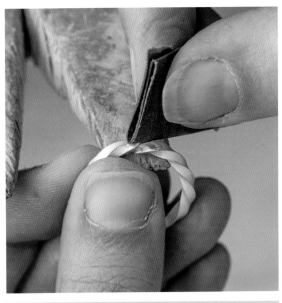

5   FINISHING_54

Core skills are used consistently throughout jewellery making. Here they are shown separately so they are easy to consult not only when you do them for the first time, but also as you continue to improve your abilities as you go through the projects.

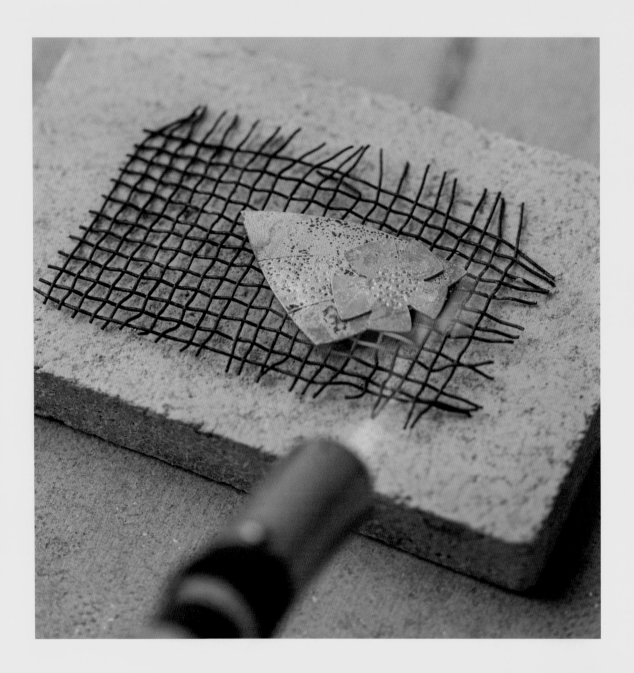

USING A PIERCING SAW

SETTING UP A PIERCING SAW

Average-size saw blades used for jewellery making are 2/0, 3/0 and 4/0 –
see the chart on page 187. The thinner the metal, or the more intricate the
design being cut, the finer the blade needs to be.

There needs to be more than one saw blade tooth to the thickness of the
metal so that the metal doesn't bump over the teeth as you are cutting, like
a bicycle going down steps.

HOW TO PUT THE SAW BLADE IN THE SAW FRAME

*Extreme close-up shot of the saw blade showing the teeth
along the front of the blade.*

1. The blade needs to be in the correct orientation so
   that the teeth are facing you and pointing down. If
   you cannot see the teeth, feel them instead. When
   you run your finger down the front of the blade
   where the teeth are, it should feel smooth. When you
   run your finger upwards along the front of the blade,
   it should feel rough. If you feel no teeth, you are
   running your fingers along the sides or the back of
   the blade where there are no teeth.

2. To hold the frame ready to put the blade in, put the
   top end of the frame in the 'V' of the benchpeg and
   the handle against your stomach so both your hands
   are free.

3. Loosen the nut a little at the top of the saw frame.
   There is now a space between the nut and the
   frame. Put the top of the blade in that space, making
   sure that that the teeth are facing out towards your
   face and down towards the handle.

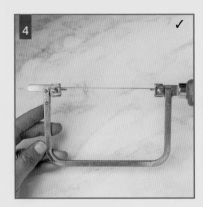 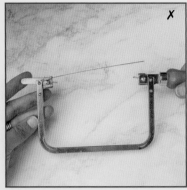 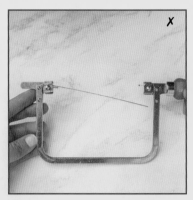

These two photographs show two incorrect ways for the blade to be positioned in the saw frame.

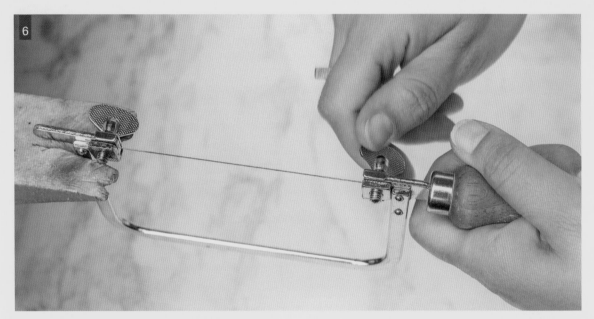

4. Tighten the top nut, making sure that the blade is positioned correctly, as shown in the photograph above, far left.

5. Place the bottom of the blade in the gap between the nut and the bottom of the frame, but do not tighten the nut.

6. Tension needs to be added to the blade. If there is no tension, the blade will not saw properly. Holding the handle, push the frame towards the benchpeg to flex the frame. While doing this, tighten the bottom nut.

Do not hold the blade when putting in the tension as it will stop tension from forming. When you release the pressure on the frame, this puts tension on the blade.

7. Test the tension on the blade. Push against the saw blade – if it is saggy and bends, there is not enough tension. Alternatively, 'ping' the blade with a fingernail. If there is a low-pitched sound, the blade is too loose. If there is a high-pitched 'ping' then there is good tension on the blade.

HOW TO USE A PIERCING SAW

Before you start to saw silver, we recommend that you practise sawing on a piece of copper sheet. Begin by sawing the metal in a straight line then progress to cutting curves and corners – see opposite page.

SAWING STRAIGHT LINES

1. Run the blade once against beeswax or candle wax to make it cut more smoothly.

2. To hold the metal, put your fingers on top pushing down onto the benchpeg with the metal spanning the 'V'. Put your thumb under the benchpeg to keep the metal stable and flat on the peg.

3. The blade cuts on the down stroke. To start cutting, gently run the blade up against the metal. This makes a small groove so that, when you cut down, the blade will stay in that groove. Cut a straight line. Keep the blade vertical. Use long strokes to use most of the blade. Do not press hard; let the blade do the work.

NOTE

When sawing, look at the shape that you are cutting, not the blade, to help you cut accurately.

HANDY HINT

If the blade is not starting the cut in the right place, use your fingernail as a guide to start the cut. Put your nail where you want the cut to be on the edge of the metal so that your nail touches the edge of the metal. This makes a corner between your nail and the edge of the metal. Place the top of the blade into this corner and saw up first to make a groove, so when you start cutting the cut is where you want it.

Another option is to start by cutting into the wood of the benchpeg at the same time as you start to cut the metal, to control where the blade starts the cut. Again, start with an upstroke to saw a groove into the metal. Once the cut has been started, move the metal back over the 'V'.

SAWING CURVES

To cut curves, keep the blade vertical and moving up and down. As the blade is going up and down, turn either the metal or the saw frame to follow the curve. If you turn too quickly, the blade will get stuck.

NOTE

It helps to realize that the cross-section of the blade is rectangular like a ruler. Place a ruler between two edges (such as between two tables) and move it up and down. Notice that it moves easily; it flows. But if you twist the ruler or change the angle from upright, the ruler catches between the edges. Now think of this as the blade travelling through metal and see how twisting or angling the frame will make the blade catch or break.

SAWING CORNERS

When you get to the corner, keep moving the blade up and down but do not move forwards (think of the motion as marching on the spot). As the blade is going up and down, slowly turn the metal until the blade is facing the new direction in which you want to cut. The blade has cut a tiny circle to spin in, allowing it to turn a corner. Proceed to cut in that direction.

For sharp corners, you can cut up to the corner from both sides rather than turning the blade at the corner. To do this, cut to the corner, then come back out along the cut you have just made. To do this, keep the blade moving up and down even though you are going backwards, as this will help the blade to move along the cut. Once you have come out of the metal, start another cut that leads to the corner. If you cannot come out of the metal to start a new cut, still come back a little along the cut you have made. Then cut across to the other side that leads to the corner and cut along that line to the corner.

SAWING INSIDE SHAPES

To cut a shape out of the inside of the sheet, not from the edge, drill a hole first. Drill the hole inside the area where the shape will go, not on the marked-out line. Undo the blade at the bottom of the saw frame. Thread it through the hole and take the metal to the top of the blade (**a**). Put the blade back in the bottom of the saw frame with tension. This way, you can saw the inside details (**b**).

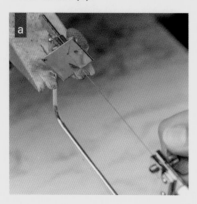

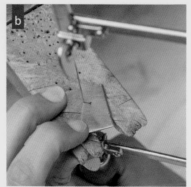

SAFETY NOTE: SAFE SAWING

When sawing, you need to hold the metal in a safe way so you do not saw into your fingers. Either hold your fingers behind where you are sawing or keep them over the wood. Do not put them in front of the blade over the gap.

Here, the fingers are to the side of the join, so that when the blade cuts through the join it will cut into the wood, not your fingers.

Here, the fingers are behind the blade, holding the ring on either side of the join. When the blade goes through it will hit the wood, not the other side of the ring.

TROUBLESHOOTING

REASONS YOUR SAW BLADE MAY BREAK:

- The metal is coming up with the blade. This is because you are not holding the metal down with your thumb under the benchpeg to keep it stable.

- You are pushing the saw blade too hard. This puts too much pressure on the blade, making it bend and break. Let the blade do the cutting – there is no need to push it hard.

- You are wriggling the blade when it gets stuck. If the blade gets stuck, you need to align the blade with the groove of the cut. If the metal is small, let go of the metal and the blade will twist the metal back into place. If the metal is too heavy for this, put gentle upward pressure on the blade and turn the frame slowly in an arch around the blade from left to right. When the blade aligns with the cut, it will free up.

IF YOUR SAW BLADE DOES NOT HAVE ENOUGH TENSION ON IT:

- Put the blade in again, making sure that the frame is flexed when tightening the bottom nut to get tension on the blade.

- If, when tightening the nut, it does not hold the blade firmly and keeps slipping, there may be a broken bit of blade stuck between the plate and the saw frame impeding the closure. Remove it and try again.

- When the saw frame gets old, you may find that the blade slips when you tighten up the nut. One way to fix this is to completely unscrew the little square plate to take it off and then put it back on again the other way around. This will make the screw tighten on a different part of the thread, not the part that is worn from usage.

FILING

Files come in a variety of shapes and usually five different cuts, which is the coarseness of the file. The higher the number describing the **cut**, the finer the file will be.

One good-quality, medium cut 2, 150mm (6in) half-round hand file and a set of needle files – also cut 2 – will set you up for most jobs.

A selection of files

Top right, a selection of riffler files; bottom, from left to right: a selection of needle files, a selection of diamond needle files and a hand file with wooden handle.

HAND FILES

FITTING A HANDLE TO A HAND FILE

Hand files do not come with a handle, so it is important to buy and fit one. The tang is the pointed bit of the file that goes into the handle. It can be dangerous if your hand slips when filing and your file does not have a handle on it.

The tang

1. To fit a wooden handle, put the file in a vice, protected with a cloth, with the tang poking out and upwards.

2. Place the starter hole of the handle over the tang. Hammer the handle straight onto the tang with a mallet. Take care to keep the handle in line with the file.

HANDY HINT

Remember that silver filings are scrap and can be collected to sell back to the bullion dealer. To catch the filings, place a sheet of paper, with a crease down the middle, on your lap.

NEEDLE FILES

Needle files perform the same function as hand files, but are smaller. Needle files are often sold in sets of six or eight files and are made of steel.

RIFFLER FILES

These are double-ended files in a variety of shapes which are the same size as needle files. They are curved, which allows for filing in awkward places.

DIAMOND FILES

These work in both directions and can file metal quickly. Because they are diamond, they will damage all stones. Steel files sit just below 7 on the Mohs scale of hardness (see pages 190–191), so they can be used to file near stones that are harder than 7.5.

CARE OF YOUR FILES

Store your files separately so that they do not rub up against each other, as this blunts them. Needle files usually come in a pack that has separate compartments.

When the file starts getting clogged with filings, use a file brush to clean it. Run the file brush on the diagonal to clear the teeth – see below.

USING FILES

- **Choose the biggest file for the job.** If you are filing an edge flat, a smaller file can drop into the dips and make them bigger. A bigger file will span the dips, taking them all down to the same height and, therefore, making the edge flat and even.

- **Files only work on the forward stroke.** You can file backwards and forwards, but not only backwards. Ease the pressure off the file when returning the stroke to prolong the life of the file.

- **Always support the work when filing.** Do not file in mid-air as the filing will not be accurate or efficient.

- When holding work against the benchpeg, have only a little bit poking up above it, to enable accurate filing and avoid bending the work.

- When **filing a shape**, saw or cut outside your scribed line and file down to it for the most accurate results.

- When **filing a long, straight edge**, the work can be supported in a bench vice with soft jaws to make it easier.

- When **filing an inside angle or corner**, file away from the angle to avoid filing a return on the angle or corner, as shown below.

- Use the curved side of the half-round file for concave curves. Use the flat side for convex curves.

Note: for filing steel, use a separate file from the one you use on silver. Don't use your good file on steel!

SOLDERING

Soldering is used to join together the different components of your jewellery. There are three main things to remember when soldering:

One: Solder does not fill a gap;
Two: Solder does not flow on dirty metal;
Three: It is the heat from the metal that melts the solder, not the heat from the flame.

Any soldering problems you encounter can be traced back to these three statements.

UNDERSTANDING SOLDER

Silver solder comes in different silver **alloys**, which melt at different temperatures. The most commonly used grades are hard, medium and easy. Note that these terms do not refer to how difficult the solders are to use but they refer to the melting temperature of each grade:

· **Hard:** highest melting temperature;

· **Medium:** medium melting temperature;

· **Easy:** lowest melting temperature.

The different grades of solder are sold in different widths in the UK, so you can instantly see which is which. Hard solder is the widest strip, medium is the thinnest and easy is the medium width: see overleaf.

It is important not to mix up the different grades of solder: once you have cut strip solder into pieces (see page 43), you will not be able to tell which grade it is. Keep the different grades of cut solder in separate, labelled containers once cut.

One method to help tell solders apart once they have been cut is to colour the strips before cutting, using permanent markers. We suggest that you colour both sides of a strip of hard solder with red permanent marker, as red is a 'hot' colour and thus associated with the solder with the highest melting temperature; likewise, blue is a cold colour, so you can colour a strip of easy solder with blue marker for the lowest melting temperature, leaving a neutral green with which you can colour medium solder. The ink will not affect the flow of the solder and will disappear once the solder is heated.

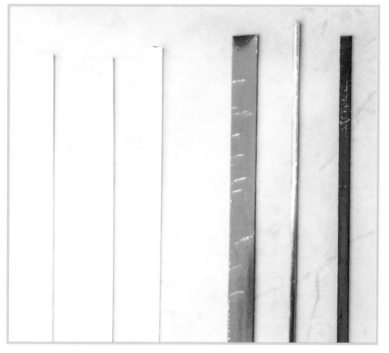

Strips of silver solder.

Strips of solder before and after colouring with permanent marker. Red is hard solder, green is medium and blue is easy.

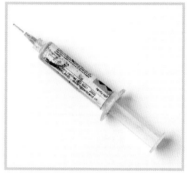

A syringe of paste solder.

In the US, solder is sold in thin wire form, which can be bent into different shapes at the end of each length so you can tell which grade is which. Solder is also available in paste form, which comes in syringes (see right). Some jewellers prefer the paste form. We use strip solder throughout this book, but you can try paste solder if you choose.

CHOOSING YOUR SOLDER

Start with hard solder for all joins that are not close to each other. Move on to medium solder and then easy solder for the last joins, such as adding a jump ring after having made a pendant. Medium and easy solder are also used for joins that are close to other solder joins. The different melting temperatures will prevent reflowing the solder joins that are already done.

CUTTING STRIP SOLDER

Solder needs to be cut into small pieces so it is easy to place and control.

1. To cut strip solder, use tin snips to cut up the strip into ribbons.

2. Use flat pliers to straighten out the ribbons, then cut across the ribbons with the tin snips. Cut the solder over a container to stop the pieces flying everywhere. The solder pieces are called **pallions**.

If you have difficulty cutting the solder, try rolling the solder thinner using a rolling mill, or hammering it using a metal hammer on a steel block. Special solder-cutting pliers are also available if you find it hard to cut solder with tin snips.

Medium solder and the US-style wire solders can be cut with wire cutters. You can also buy pre-cut solder.

CHOOSING YOUR SOLDERING SURFACE

There is a variety of soldering surfaces available: the three used in this book are soldering blocks (asbestos substitute), charcoal blocks and steel mesh. These three together cover most soldering needs.

It is worth noting that charcoal absorbs oxygen, so fewer oxides form on the metal, creating a cleaner soldering environment. You can carve into one side of the surface to create a dip that prevents the molten metal running off the block. Charcoal blocks can, however, crack, so to prolong the life of the block you can gently heat your block before using it for soldering. You can also bind the charcoal block around the sides with binding wire (as below) or copper wire to hold it together.

Another option is to make a tray for your charcoal block out of thin steel, brass or copper – see overleaf. This tray will keep the block intact and prolong its use.

Try to avoid heating the edges of the block, as they burn away more quickly.

All three soldering surfaces should be placed on a larger fireproof surface such as a fireproof soldering sheet, as shown on pages 18 and 28. Throughout the book, you can use a soldering block or a charcoal block.

Soldering surfaces.

a) Soldering block; b) steel mesh and c) charcoal block.

MAKING A METAL TRAY FOR A CHARCOAL BLOCK

1. With a permanent marker, draw around your charcoal block onto the sheet of metal you are using to make the tray. Add no more than half the depth of the block on all sides. You can use thin steel, brass or copper for this. Draw diagonals across the corners.

2. Cut out the tray template from the sheet metal using tin snips.

3. Use a vice to bend up the sides. Put one short edge in the vice with the jaws 1mm (1⁄16in) outside the mark line (this allows for the thickness of the metal when you bend it up, otherwise it will be too small). You can use a mallet to hammer over the vice jaw. Do the same for the other short side. Then repeat on the longer sides. Move the tray along the vice and continue to bend up the sides, if the vice is narrower than the long edge of the metal sheet.

4. Place the charcoal block in the tray and squeeze the sides up to hug it.

CLEANING AND FLUXING THE JOIN

Before soldering, you will need to prepare the join by cleaning and fluxing it, or the solder will not flow. As the flux is water-based, the join must not be greasy, so must be cleaned first then fluxed. Flux keeps the join clean while soldering by preventing oxides forming. There are different types of flux and they can burn out at different temperatures: your choice of flux should go molten at the same time at which the solder runs. This will keep any join clean and help the solder flow. If the join is not fluxed, the solder will not flow there.

Borax is an inexpensive and effective flux for silver soldering.

1. Make sure that that the join area is clean. Filing and sawing cleans the join. If you then rub over it with your fingers, the join may be greasy, which will inhibit the flux and, therefore, the soldering. To degrease, sand the join area and then avoid touching it. If the join area is oxidized (tarnished), use a sanding stick (see page 55) to remove the tarnish. Remember to sand the surfaces coming together in the join.

2. Flux the join. If you are using a borax cone, grind the cone in the borax dish or on slate with water until you have a white liquid the consistency of skimmed milk. If the flux is too thick, it can bubble up when heated and dislodge the solder.

3. Use a paintbrush to paint flux over the join. Fluxing can be done on the block or while holding the piece. The flux keeps the join clean during soldering by stopping oxides from forming.

4. The flux is water-based, so if the join area is still greasy the flux will ball up like mercury over the surface. Conversely, if the join is too clean, the flux can also ball up. Either way, you will need to sand over the piece again to remedy this. When it is prepared correctly, the borax will coat the join.

HANDY HINT

If you flux your piece on – rather than over – the soldering block, flux can start to collect on the block and, consequently, your work can stick to the flux on the block. Then when the piece is moved after soldering it can take pieces of the block with it.

If the block becomes pitted because of this, it can be resurfaced by grinding it on a brick or on pavement. Alternatively, if you have two blocks, they can be rubbed against each other to make the surfaces flat again.

SOLDERING YOUR WORK

TYPES OF SOLDERING

There are different ways of placing the solder onto a join. Depending on the method you choose, they have different names:

1. Pallion or panel soldering This is the primary method of applying the solder that we have used in the projects in this book. Cut small pieces, called pallions or panels, of solder and place them on the join – this is demonstrated below.

2. Pick soldering This is where a piece of solder is heated into a ball and is picked up on a steel pick or tweezers. It is then placed on the join while heating the work, rather than placing it on the join while all the pieces are cold.

Pick soldering is explained in project 3 – Stone-chip earrings – on page 96.

3. Sweat soldering This method is where the solder is pre-flowed on one piece – the pieces are put together and then the solder is reflowed. This technique is explained in project 6 – Brooch with cut card technique – on pages 152–154.

4. Stick soldering This is where the solder is melted directly from the stick or wire onto the work; the solder is not pre-cut or pre-flowed. This method is more common whenever you buy solder in wire form, as in the US, or in silversmithing. We have not used this method for the projects in this book.

PLACING SOLDER ON THE JOIN

The size of the piece of solder needed is something you learn with experience. If it is too big, there will be a blob that needs to be filed off later. If it is too small, the solder will not fill the join and more solder will need to be added.

Use a paintbrush dipped in flux to pick up the pallion and place it on the join.

The solder can go on top, under or against the join: it must be touching the join. If you are soldering a smaller piece of metal to a bigger piece of metal, place the solder on the bigger piece against the smaller piece. With the solder on the bigger piece (which will take longer to heat), you have less risk of the solder flowing onto the smaller piece first.

Use tweezers or a paintbrush that has been dipped in flux to pick up the pallion of solder. Place the pallion on the join. It is easier to do this if the flux on the join is wet: if the flux is dry, the pallion may fall off.

From left to right: the solder pallion placed on top, against the front, against the back and under the join.

SUPPORTING YOUR WORK ON THE SOLDERING SURFACES

Sometimes you need to hold your work in place for soldering rather than simply rest it on the soldering surface. To start with, here are some good options to try:

Tweezers These can be used to prop up the work from the soldering surface, as in the stone setting project (see page 133). If you need to stop a piece moving or rolling away, tweezers can be placed next to it to hold it in place during soldering. Remember they get hot, so don't pick them up afterwards.

Reverse-action tweezers Some pieces need to be held in place, such as earring posts. Reverse-action tweezers are perfect for this. Be careful when using reverse-action tweezers: their grip is very strong. If you hold them across a gap, or hold a hollow piece of metal, they can crush the pieces during soldering. It is best to hold pieces in the tip of the tweezers as it is easier to hold the piece there and the tweezers do not get in the way of heating the join.

Binding wire You can wrap binding wire around a piece to bring the join together. Always remember to remove the binding wire before pickling.

Holding an earring post in place for soldering on the block using reverse-action tweezers.

UNDERSTANDING THE SOLDERING TORCH

Most crème brûlée torches have built-in ignition systems. If you are using a torch without this function, place a lighter under the nozzle of the torch, when the gas is on, to light the torch. If you place the lighter too far in front of the nozzle, it will not light the gas. If you place it too close to the blast of the gas, the gas will blow out the lighter before the gas lights. The lighter must be placed right under the end of the nozzle.

Place the lighter below the front of the nozzle to light a torch.

ABOUT THE FLAME

The flame has two parts – the outer blue cone and the inside, lighter blue cone. The hottest part of the flame is just in front of the lighter blue cone. If the nozzle is too close to the piece that is being soldered, you will hear a whooshing sound, which will indicate that you are not in the hottest part of the flame. Move the flame back until the whooshing sound stops.

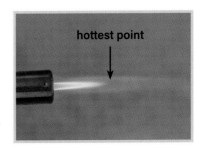

Parts of the flame: the inner and outer blue cones.

ANGLING THE FLAME

The angle at which you hold the flame will affect how your silver piece heats. Hold the flame at the angle at which you would hold a pencil to write, at about 45 degrees. The lower you hold the flame, the more of your work is in the flame. The steeper the angle, the smaller amount of your work is in the flame. Use this information to adjust your flame for different heating needs as necessary. It might help to imagine the footprint of the flame like a spotlight on a stage. The lower the spotlight, the bigger the light is on the stage and the higher the spotlight, the smaller the light is on the stage.

Try pointing the flame at the fireproof block and move closer until you are at the hottest part of the flame (at the end of the lighter blue cone). You will see a bright red dot. Now move closer. There will be a whooshing sound and the bright red dot will have a black dot in the middle where the flame is cooler.

Hold the flame at the angle at which you would hold a pencil.

ESSENTIAL SAFETY ADVICE

- Keep the flame up and forwards when it is lit but you are not yet soldering. Do not point it down thinking it is safe – up is safest as it will heat nothing but the air in front of you.

- Do not point the flame at anything flammable.

- Avoid holding your flame vertically down – it will splutter as it is blowing back on itself.

This black dot shows that the flame is too close to the block.

HANDY HINT

It is important to get used to holding the torch in your non-dominant hand. This is so that you can hold the tweezers in your dominant hand when you come to flowing the solder (see opposite), which will give you much more control and dexterity with the tweezers.

FLOWING THE SOLDER

The heat of the metal, not the heat from the flame, will cause the solder to flow. This is really important to understand. Watch the colour of the flux and the silver when heating: borax flux will go brown and shiny at the point at which the solder is about to flow.

When silver is heated, it will go through a sequence of yellow, grey, pink, then dull red. If it turns bright red, it is heading towards melting point. It helps to get used to the colour changes so that you know what the silver should look like when it is at soldering temperature. If it stays grey, it is too cool and will not get hotter. You may not see all the colour stages if the light in the room is too bright or if the piece is small. It is important to keep the lighting in your work space consistent so that you get used to the colour changes you expect to see.

If you are soldering together two different-size pieces of metal, both pieces need to reach soldering temperature at the same time. The larger piece will need to be heated first, as it needs more heat than the smaller piece to get up to soldering temperature. Do not point the flame at the solder and expect it to run, because it won't run until the metal is hot enough to melt the solder.

When both pieces to be joined are hot enough and equally heated, you will see a shiny silver line along the join line, which is the solder flowing. Remove the flame when this occurs. Do not linger with the flame on the piece, as it will quickly get to the melting point of the silver.

When silver solder reaches its melting temperature, it becomes fluid and 'flows' into the join, and bonds the two sides together. This is known as 'hard soldering' or 'brazing', because the melting temperatures are above 430°C (806°F), but commonly called 'soldering' in jewellery making.

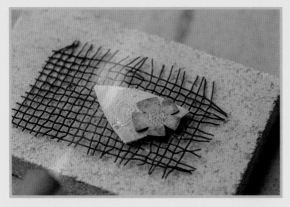

You can see that the flower is still dark grey, indicating that it is not yet up to soldering temperature. Another indicator that the piece is not yet up to soldering temperature is that the borax on the triangle is still white, not shiny brown.

Both pieces in this photograph, even though they are different sizes, have reached soldering temperature at the same time. You can see this because they are the same colour.

Once soldered, you will see that your piece looks 'dirty', with grey areas and bits of glassy borax. The next step will be to clean this off in the pickle – see pages 52–53.

TROUBLESHOOTING

IDENTIFYING SOLDER PROBLEMS

PROBLEM

The solder has not flowed: it has balled up on the join and not run.

SOLUTIONS

- The join could be dirty with grease or oxides, or it wasn't fluxed, so needs to be cleaned. Try putting more borax on by placing the borax cone directly on the join while it is hot (so that you don't burn the paintbrush) and try to solder again. If this does not work, pickle and rinse to clean (see pages 52–53) and then re-flux and try again.

- It could be that the join area didn't get hot enough. Refer to the information on page 49 (Flowing the Solder) about the colours you should expect to see during the heating process. Try again, making sure that that you heat the work, not just the solder.

PROBLEM

The solder has flowed but not filled the whole join. You will see this as a black line which is a shadow along the join where the solder has not filled the join.

SOLUTIONS

- Not enough solder was used. Add a small piece of solder on the join and heat again. (This was the case in the photograph above.)

- If the black line is on the underside, the side that was on the block was cooler, so did not reach soldering temperature. Turn the piece over so that part of the join is uppermost and heat up again to draw the solder through the whole join.

PROBLEM

The solder flowed, but not on the join.

SOLUTIONS

- One side got hotter than the other, so the solder flowed towards the heat instead of down the join. Add another small piece of solder over the join, which will help the solder that's already there to flow into the join. Adding another piece of solder will break the surface tension and help the solder that's already there to flow into the join. Heat again to reflow, making sure that both sides of the join come up to soldering temperature at the same time.

- The join wasn't touching. Solder will not fill a gap. Pickle and rinse to clean. Correct the join (see Troubleshooting, page 67). Flux, put on a new piece of solder and solder again.

- The solder was not placed properly and was not touching the join. Put a small piece of solder on the join and solder again.

- The join opened while heating. **Quench** (cool the piece in water), correct the join, re-flux, put a small piece of solder on the join and solder again. This problem can be prevented by **annealing** a piece (see page 78) before soldering so any tensions in the piece are released, thus it does not move during soldering.

- If there is already quite a lot of solder on the join but it has not run into the join, you can move the solder with tweezers as it is flowing – it is like spreading butter with a knife. Push the solder over the join and **capillary action** will suck it into the join. Capillary action describes the ability of a liquid to flow against gravity in a narrow space – in this case, the solder will flow into the join.

PROBLEM

A part of the piece melted before the solder flowed.

SOLUTION

The hottest part of the flame was on the wrong area. If it cannot be disguised, you will need to start again.

PROBLEM

The solder has balled up, but is not on the join.

SOLUTION

Keep the heat on the piece and move the solder with tweezers. The heat will unstick the solder from the flux and allow you to move it with the tweezers. Do not move the flame away and then try to move the solder, as it will either not move or it will fall off. Make sure that you do not point the flame at your hand.

PROBLEM

After soldering, there is pitting around the join. This happens when the piece has been overheated and the solder has alloyed with the silver, reducing the silver melting point. This creates pitting or indenting where the solder was placed.

SOLUTIONS

- If there is enough thickness of metal on the piece, you can file and sand out the marks.

- If there is not enough metal to file and sand away the marks, you can flow a lower melting temperature solder over the pitted area to fill them. Then file and sand to clean.

- If you cannot remedy the pitting with the two solutions above, you will need to start again.

PICKLING

After soldering, the piece needs to be put into pickle to clean off the oxides and flux – you should quench (cool) the piece in a bowl of cold water before pickling, so that the pickle doesn't splash back on you.

HEATING AND USING THE PICKLE

Pickle works much better when it is warm. You can put the work in when the pickle is cold, but it will not clean well until the pickle has heated up. To keep the pickle warm (not boiling!), use a slow cooker, as we demonstrate throughout this book.

1. Mix your pickling powder with water in a glass bowl that fits inside your slow cooker. Follow the manufacturer's instructions on the ratio of pickle to water.

2. Pour water into the slow cooker until it is approximately half full.

3. Place the bowl with the pickle solution into the water in the slow cooker.

4. Set the slow cooker on low to heat the water and, therefore, your pickle.

5. Keep the slow cooker covered with its lid to prevent too much evaporation of the water, and add more water if the level becomes too low. This applies to both the water and the pickle solution.

6. Place your piece to be pickled in the solution using tweezers. If you use steel tweezers, be sure that they do not touch the pickle (see opposite page).

7. Leave the piece in the pickle until it is clean. You will know that the silver is clean when it turns white. This usually takes no more than a minute or two, depending on how new and how hot the pickle is. If you have added water repeatedly to your pickle and it has become diluted and is not working as well, top it up with approximately half a teaspoon of pickling powder.

At some point you will need to replace the pickle – see opposite for guidance on how to dispose of the old pickle safely.

REMOVING YOUR PIECE FROM THE PICKLE

Use brass, plastic or copper tweezers or tongs to remove the piece from the pickle. You should not use steel implements to remove your work from the pickle as steel creates an electrical circuit that will copper-coat the work in the pickle – see below, right. If this happens, however, you do not need to throw away the pickle or the piece itself: you can remove the copper coating by gently heating your work and pickling it again (usually two or three times) until the coating is gone. If the piece is not intricate, it can be sanded off.

Rinse both the piece and the tweezers in water. It is good practice to rinse the tweezers, otherwise you will get acid on your work surface; in addition, metal tweezers will corrode over time.

Here, the ring in the pickle is clean, ready to be removed. The other ring is being put in with brass tongs.

This photograph shows what a silver ring looks like when it is copper-plated in the pickle.

DISPOSING OF YOUR PICKLE

Safety pickle will work for quite a long time without needing to be replaced. When it does need to be replaced it will start turning dark blue, and will simply stop cleaning your metal pieces.

To dispose of the used pickle, add bicarbonate of soda (baking soda) to cold pickle until the liquid stops fizzing. It will bubble up, so make sure that your container allows for this, or put it in a sink. Dispose of it safely down a drain.

FINISHING

Finishing involves removing the deepest marks with increasingly finer marks until the marks are so fine that the surface looks smooth. Deeper marks are removed with either files or abrasive paper. Do not go too rough because you are adding unnecessary marks and, therefore, more work. Do not go too fine because it will take longer to remove the marks.

The principle is to use an abrasive that is just rough enough to remove the marks. Don't move on to a finer abrasive until you can see you have removed all the marks at that stage.

You can use abrasive paper on its own, but we recommend using an abrasive sanding stick. You can make your own sanding stick following the instructions on the opposite page.

POLISHING

Polishing is the finest stage of abrasion, to make the surface as smooth as possible. Imagine lights rays as a bouncing ball. When you bounce a ball on a smooth surface, the bounce is uniform. When you bounce a ball on an edge, it bounces off in different directions. So when light rays hit a smooth surface they bounce off (reflect) in a uniform way, creating a shine. If the light rays hit an edge of a scratch they bounce off at random, so it doesn't look shiny because the reflections are not uniform.

There are many ways to polish. Here, we demonstrate three methods that will cover all the projects in this book. We have chosen these methods as they work well, cover different budgets and are convenient and safe.

BARREL POLISHERS

A barrel polisher tumbles the work with steel shot in a barrel to burnish the jewellery and impart a shine. Depending on the barrel polisher, it will take fifteen to thirty minutes to polish a small piece of jewellery. Some barrel polishers have a timer, so you can leave it to run and it will switch itself off when it has finished the cycle. It will not remove any scratches, so if the sanding is not good, it will not improve it. On some models of barrel polisher, the lids push fit – such as on rubber barrel polishers; on other models, the lids latch into place and are, thus, more secure. In all cases, the lid and the whole area around the lid must be dry before you close the barrel or the polisher will leak. Instructions for using a barrel polisher can be found on page 56.

Steel shot can be purchased either in ball form, or in a variety of different shapes, which will reach more areas of your work. Store the shot in the liquid in the barrel to stop it from rusting.

Barrelling compound is a mixture of soap and rust inhibitor that makes the barrel polisher work more efficiently.

MAKING AN ABRASIVE SANDING STICK

1. Choose a grit of abrasive paper.

2. Choose a piece of wood (this can be a piece of dowel rod) that is longer than the piece of abrasive paper is when laid lengthways.

3. Stick a length of masking tape on the abrasive side of the paper along both of long edges. The masking tape should be longer than the width of the abrasive paper and should hang over the edges of the paper.

4. Turn the paper over.

5. Line up the stick with the edge of the abrasive paper so that the masking tape is at the top and bottom of the stick.

6. Attach the end of the masking tape around the stick.

7. Score the side of the abrasive paper down the side of the stick with a scriber, taking care not to go all the way through. This is to give the paper a crisp edge as you wrap it around the stick.

8. Score every time you turn the stick. As you turn, pull the paper in order to keep it tight against the stick. Keep pushing the masking tape down on the stick as you go.

9. At the end, fasten the tape around the ends of the stick to hold the paper in place. Write the grit grade on the masking tape.

As you use the sanding stick, the abrasive paper will wear out. Tear off the worn sections and use the fresh paper underneath.

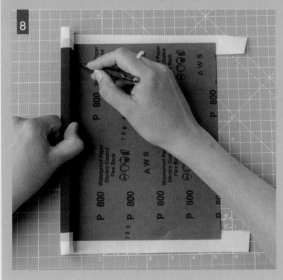

USING A BARREL POLISHER

HANDY HINT

If you are using the model of barrel polisher shown, pour the contents out between the chrome bars to avoid getting any shot stuck inside the hole that holds the lid.

1. Fill the barrel with steel shot, to no more than a quarter of the barrel.

2. Add water to no more than halfway up the barrel. Into that, add a heaped teaspoon of barrelling compound. Do not overfill the barrel, or it will not turn properly.

3. Put your work into the barrel.

4. Make sure the lid and the top of the barrel are dry and secure the lid in place.

5. Load the barrel onto the base and turn it on. If it has a choice of speeds, chose the lowest speed.

6. After 15 to 30 minutes, pour out the steel shot and polished work into a sieve that is over a jug. Move the sieve around, as if panning for gold, and your work should come to the surface. Take the work out, while wearing gloves, and rinse it in water. Pour the shot from the sieve into the jug, then pour all the contents of the jug back into the barrel.

SAFETY NOTE

Use gloves when putting your hands into the liquid at step 6.

Once you have taken the work out of the barrel polisher, it will need to be rinsed in water. Dry your work well after removing it from the water to avoid getting water marks on the silver.

Where the shot doesn't reach, you will see a white outline or halo (see right). If the shot you have chosen leaves a white outline, use a brass brush to burnish your work before putting it in the barrel polisher. Alternatively, you may like to keep this as a design feature.

POLISHING BY HAND

If you don't have a barrel polisher, polishing by hand is a good option. Here are two methods for polishing by hand:

1. a) Use metal polishing wadding (such as Brasso) to rub vigorously over your piece.

b) Rub with kitchen paper to bring up to a shine. It is a good idea to use gloves to do this, as your hands will get very dirty.

Alternatively:

2. Run a brass brush backwards and forwards over the piece. Do not press too hard, to avoid impregnating brass into the silver; alternatively, dip the brush in water before you polish your piece or do it under running water. You can add a bit of liquid soap to help lubricate the brush.

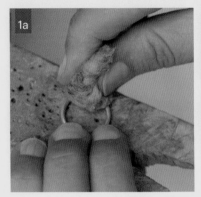
Putting on polish with the wadding.

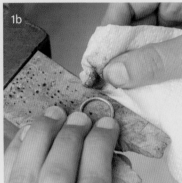
Taking off the polish to bring up a shine with kitchen paper.

HIGHLIGHTING AFTER POLISHING

After polishing, you can go on to burnish the corners of your piece. Burnishing compresses the surface of the metal to impart a high shine. It works best on corners, edges and high spots. You can **burnish** using the back of a steel teaspoon or a burnishing tool. Here, we are using burnishing to highlight the edges of a square-wire ring.

After brass brushing, rub quite hard with the back of the bowl of the teaspoon over the edges of the work.

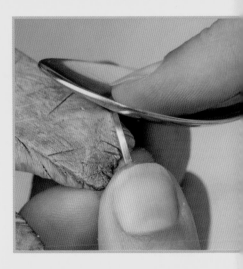

THE PROJECTS

1 TWO SIMPLE RINGS_60

2 TWO TWIST RINGS_76

3 STONE-CHIP EARRINGS_92

4 TEXTURED PENDANT_112

5 STONE SETTING FOR A RING_126

6 BROOCH WITH CUT CARD TECHNIQUE_146

7 FUSED PENDANT AND STUD EARRINGS_162

PROJECT 1:
TWO SIMPLE RINGS

The finished rings in square and round wire.

WHAT WE ARE MAKING

Two silver rings, one from square wire and one from round wire.

Making these two rings will introduce you to the core skills necessary for making any piece of jewellery. Right from the start, you will be making items that look great and are fun to wear.

CORE SKILLS

Sawing wire; filing; forming; soldering; sanding; pickling; polishing; burnishing.

NEW SKILLS

Sizing rings; forming rings.

YOU WILL NEED

BENCH, HAND AND FORMING TOOLS

- Benchpeg and anvil (required for every project)
- Ring sizers
- Ring mandrel
- Piercing saw and saw blades
- Beeswax or candle wax
- Half-round pliers
- Flat pliers
- Hand file
- Scriber
- Hide mallet

SOLDERING TOOLS

- Flux (such as borax) and dish or piece of slate
- Large soldering sheet
- Soldering block or charcoal block
- Paintbrush
- Torch for soldering
- Steel tweezers
- Tin snips
- Pickle in slow cooker
- Glass bowl with water and tongs (copper, brass or plastic)

FINISHING AND POLISHING TOOLS

- Abrasive paper
- Sanding stick (optional)
- Barrel polisher/brass brush or brass polish wadding (Brasso)
- Burnisher or teaspoon

MISCELLANEOUS ITEMS

- Thin marker pen
- Two strips of paper

MATERIALS

- 75mm (3in) of 1.5mm (15-gauge) square sterling silver wire
- 75mm (3in) of 2mm (12-gauge) round sterling silver wire

The lengths listed will make rings of most sizes except the very largest – i.e. above UK size V, US size 10¾. Add 10mm (⅜in) to the length of the wire for above-average ring sizes.

- Hard solder

SIZING THE RINGS

Follow the instructions below to work out the length of the wires to be cut to make the round- and square-wire rings. Follow the steps once for the round-wire ring and once for the square-wire ring. Ensure that you have two strips of paper ready.

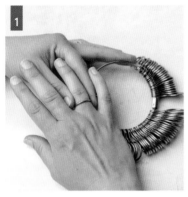

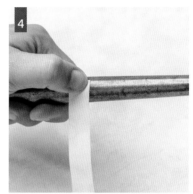

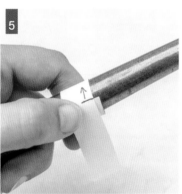

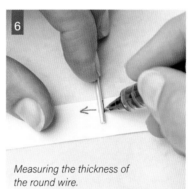

Measuring the thickness of the round wire.

1. Work out your ring size by measuring your finger with ring sizers. Your size will be the one that is tight enough that it takes a wiggling motion to remove the ring over the knuckle.

2. Slide the ring sizer that is your size onto a ring mandrel until it stops.

3. Put your thumbnail against the ring sizer at the widest side of the ring mandrel to mark the spot.

4. Leaving your thumb in place, remove the ring sizers from the mandrel. Hold a strip of paper in place with that thumb and wrap it around the mandrel.

5. Make a pen mark on the piece of paper where it overlaps. Draw an arrow pointing to the side of the paper that was wrapped around the mandrel, so that you don't refer to the wrong side of the paper.

6. Add the thickness (not the width) of the silver wire you are using to the length on the piece of paper to ensure that the inside of the ring will be the right size, not the outside – otherwise your ring will be too small. Measure the thickness of your wire with a ruler or by holding the wire on the paper next to the mark you made and then drawing another line to add the metal thickness. Round wire will add 2mm to the length and square wire will add 1.5mm to the length – these are the thicknesses of the wire being used in this project (see page 61).

Repeat these steps so that you have two measurements – one for the round-wire ring and one for the square-wire ring. On page 64, you will transfer these measurements to the wire and cut them to length.

PREPARING TO SAW THE WIRES

FILING THE ENDS

Before sawing the wires to the right length for your rings, you need to file one end of each wire. The ends of the wire must be at right angles to the edges. When you buy wire, the ends may not be straight or, if wire cutters have been used to cut the wire, the ends may be pinched. Correct these by filing one end of each wire square.

File the ends using the flat side of a hand file. Remember the following tips:

- Files only cut on the forwards motion. You can draw the file back on the metal and file forwards, or lift off the metal and file forwards, but do not only file backwards.

- Always support the metal against the benchpeg when filing. Filing without support means both the metal and the file will move around and the result will not be flat.

- It is easy to file away the corners more than the middle of the end by accident, creating a curve on the end of the wire. Be aware of this and file straight to avoid this. It is important to get the end straight, otherwise there will be gaps in your join when you come to solder.

GOOD ENDS VERSUS BAD ENDS

Looking straight at the end of the wire, you should see the shape of that wire. On the end of the round wire, you should see a flat circle. On the end of the square wire, you should see a flat square. Imagine that you are going to make a print with the end of the wire – that is how flat it should be.

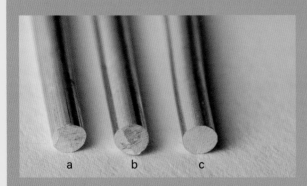

Round wire – two bad ends, one good

a) has not been filed flat and the edges have been rounded off; b) has two clear facets at the end; c) is a good, flat end showing a clear, visible circle.

Square wire – one good end, two bad

a) is a good, flat end showing a clear square; b) has not been filed straight – the corners have been filed away and there is no clear square at the end; c) the end has been pinched by wire cutters.

MARKING THE WIRE FOR SAWING

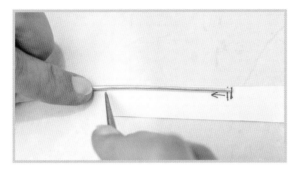

Having filed one end of each wire flat, mark each length of wire with a scriber using the length measurements you made on your pieces of paper (see page 62, Sizing the Rings). Use a scriber rather than a pen, as you cannot rub out the marks made by a scriber with your fingers and it marks accurately, whereas a pen line is quite thick and inaccurate.

SAWING THE WIRES FOR THE RINGS

Hold down the first of your two lengths of wire on the benchpeg so that it doesn't move around – the photographs below show the square wire being sawn. Hold it with your middle finger and thumb stabilized against the side of the benchpeg and your index finger holding down the wire itself.

Use the nail of your index finger to guide the start of the cut.

HANDY HINT

Turn to pages 34–38 to learn how to set up your piercing saw, and practise sawing on a piece of copper before cutting the wires in this project.

1. Cut through the wire slightly outside where you marked it. This is so this end can be filed flat without making the wire too short. The difference between ring sizes is 1mm (approximately 1/16in), and it is easy to file away a millimetre and end up making the ring too small.

2. It is good not to have a **burr** at the end of the cut. A burr is a bit of metal left at the edge which sticks out. To avoid this, cut the wire most of the way through and then push it to the end of the 'V' in the benchpeg. Saw through the wood and wire at the same time for the last bit of sawing. Also, as you near the end of the cut, the wire can close on the blade, trapping it. If this happens, push the cut open and continue to saw.

3. If the end of the wire is not straight or has a burr on it, file it.

4. Repeat the steps with the second length of wire. Remember to keep the bits of silver that you cut off and any other scrap that you end up with from the following projects, to use in project 7 for the fused pendant and stud earrings (see pages 162–175).

Your wires are now ready to form into rings.

STARTING THE RINGS

THE ROUND-WIRE RING

A. BENDING THE WIRE

1. Grip one end of the round wire with half-round pliers. Hold it so that the half-round side of the pliers is away from you.

2. Bend the end of the round wire around the half-round nose of the pliers.

3. Instead of continuing along the wire, move to the other end of the wire and make the same curve. It is easier to bend the ends when you have a long lever. Continue to bend the wire along its length around the half-round nose of the pliers until the two ends are close.

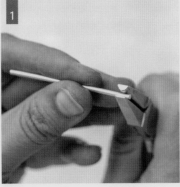
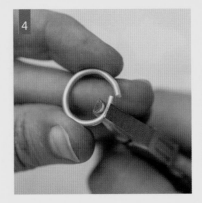
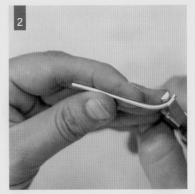
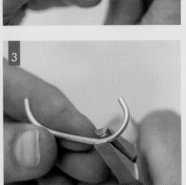

> **HANDY HINT**
>
> *When bending the wire, keep in mind the curve of a ring so that you don't overbend the wire, making the curve too tight.*

B. MAKING A GOOD JOIN

Now that the ring is nearly formed, line up the join.

4. Once the ends are close to each other, push them past each other a bit and gently pull back. This creates tension that makes the two sides sit snugly against each other. It does not matter if the ring is not round (**forming** happens later). Make sure that the ends are lined up from all sides.

5. If the join is good, there should be no gap to see light through when the ring is held up.

On the following page, we show you how to form the square-wire ring in the same way.

THE SQUARE-WIRE RING

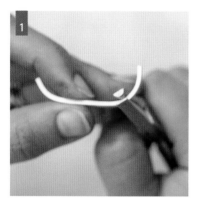

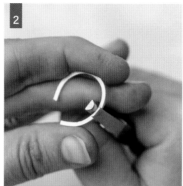

1. As with the round wire on page 65, bend both ends of the square wire into a curve around the half-round nose of the pliers.

2. Move along the wire with the pliers to create a ring shape. Square wire needs extra attention because it can twist. If you work along the wire a bit at a time with the pliers, it will avoid a twist happening.

3. Line up the join for soldering. Make sure that all corners line up and that you have a good join – see opposite, Troubleshooting: Correcting Bad Joins.

The round- and square-wire rings are now ready for soldering (see pages 68-69).

TROUBLESHOOTING

CORRECTING BAD JOINS

PROBLEM

A 'top V': the ends are meeting at the bottom but not at the top.

SOLUTION

Use half-round pliers to move both ends down, one at a time. This will close the gap.

PROBLEM

The ends are not filed flat and meet only in the middle.

SOLUTION

To correct, put the ring on the benchpeg as in the right-hand photograph on page 38, Safe Sawing. Cut through the join with the saw. The saw will act as a file on both sides at the same time, bringing the join together. The join needs to be touching for this to work. Alternatively, twist open the ring and file both ends.

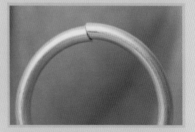

PROBLEM

A 'side V': although the ends are flat, they have been filed at an angle so there is a gap between them.

SOLUTION

Twist the ring to open it sideways so that you can reach each end with a file. File the ends flat and reset the join.

PROBLEM

The ends are not close enough together to solder.

SOLUTION

Twist the ring to open it sideways so that you can push the ends slightly past each other. There needs to be some tension between the two sides so that they sit against each other. Pull back the ends to line them up – you will hear a little click when they line up.

PROBLEM

The ends are not aligned: one end is higher than the other – they need to line up.

SOLUTION

To correct, use half-round pliers to align the join.

SOLDERING AND PICKLING THE RINGS

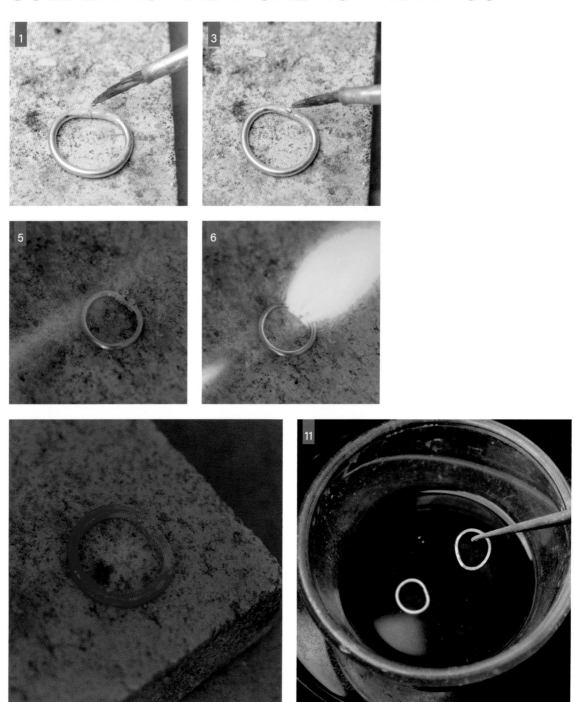

The square-wire ring after the solder has run.

1. Paint flux on the join of both the rings. If the flux is balling up, the join is dirty. Clean it (see page 45) and flux again.

2. Place the first (round-wire) ring on a soldering block with the solder join facing away from you. This way, there is less chance of melting the side of the ring opposite the solder join.

3. Place a piece of hard solder on the ring, touching the join. It can be under, against or on top of the join, as long as it is touching the join. Always use hard solder for your ring joins, in case you want to add more components to your ring later.

4. Hold the torch in your non-dominant hand. Light the torch. Hold the tweezers in your dominant hand. This allows you better control over the tweezers. It also reduces the risk that you will pick up hot metal with your hands.

5. Solder. Avoid blowing the solder off the ring by dipping the flame on and off the join area until the flux has dried. When drying the flux, keep the flame off the work when there is bubbling and put it back on when the bubbling stops. The flux will become white when dry.

6. Once the flux is dry, keep the heat on the join of the ring. Remember, the heat of the metal will cause the solder to flow, not the heat of the flame. The solder will flow towards the heat. If one side is hotter, the solder will flow to the hotter area. Position the ring join in the middle of the flame at the point just past the light blue cone. There will be a bright silver flash when the solder flows. Remove the flame when this occurs by lifting up the flame. Do not linger with the flame on the ring, as it will quickly get to the melting point of the silver.

7. To be sure that the solder has flowed through the whole join, turn the ring over with tweezers. If there is a black line at the join, it means the solder has not run all the way through. If that is the case, heat the join from this side so that the heat will draw the solder to this side of the join. If you see a silver line, the solder has flowed through.

8. Turn off the torch and put it down on a fireproof surface or on a fireproof hook.

9. Repeat steps 2–8 for the second (square-wire) ring.

10. Quench (cool) the rings in water and put them in the pickle.

11. When the rings are white, the oxides and flux have been cleaned off by the pickle. Use brass, plastic or copper tweezers or tongs to remove them from the pickle. Rinse the rings and the tweezers in water.

12. Dry the rings, ready for the next stage.

HANDY HINT

It is quicker to set up both the rings to be soldered at the same time. Place them away from each other on the soldering block and solder one at a time.

FORMING THE RINGS

Form one ring at a time.

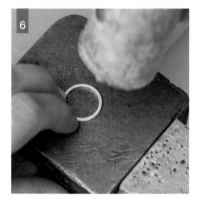

HANDY HINT

To hammer efficiently, hold the mallet at the end of the handle, not close to the head of the mallet.

1. Check the joins of your rings. Do not start forming the rings until the ring joins are soldered correctly. If you reflow the solder after forming the ring, there is a risk the join will spring apart because of the tension put in the ring by hammering. If the join needs correcting, do it now (see pages 50–51 for guidance on correcting the join after soldering).

2. Once the join is good, file off any large lumps of solder on the join before forming. This avoids hammering the solder into the silver.

3. If either ring is not round enough to fit on the mandrel, hold it upright on the table or steel block. Use a mallet to tap it, from the top, back into a round shape, rather than oval, so that it will fit on the mandrel.

4. With the ring on the mandrel, hold the ring with your index finger and thumb at the back of the mandrel. Rest the handle of the mandrel against your legs or waist. The ring needs to be held at this point or it could fly off the mandrel. Hammer with a mallet.

5. When the ring is rounder, change to holding the mandrel at the handle with the point pointing upwards. You do not need to hold the ring any more. Turn the mandrel while hammering to make the ring round.

6. If the ring does not lie flat on a flat surface, and rocks like a seesaw, it needs to be malleted flat. Hold down the ring with your fingers positioned as in the photograph. Do not hold your fingers flat – it is all too easy to hammer them by accident. Tap down the ring on a steel block to make it flat. Hammer around the whole ring, turning it as you go.

7. Check the ring sizes by trying each one on. Once the rings fit, they are ready for finishing.

ADJUSTING THE SIZE OF A RING

THE RING IS TOO BIG

If the ring is too big, cut out a piece at the solder join with the saw and solder shut again. As a guide, cutting out 1mm (approximately ¹⁄₁₆in) takes the ring down one size. Alternatively, saw through the join and then overlap the join. Try it on and adjust the overlap until the ring fits. Then saw off the overlap.

THE RING IS TOO SMALL

If the ring is too small, you will need to stretch it. To stretch the ring, put it on the mandrel. Hold the mandrel at the pointed end and rest the handle on a solid surface. Hit downwards on the ring with a mallet – see right. Rotate the mandrel as you hammer, as this will also rotate the ring. The hammering will force the ring down the taper of the mandrel, therefore stretching it. Once done, the ring will probably not be flat, so mallet it down on a steel block again as in step 6 on the opposite page. To avoid making the ring too big, keep checking the size as you go. With the square-wire ring, turn it over to stretch both edges in the same way, otherwise the ring will be tapered.

If the ring is too small, another option to stretch it is to use a steel hammer to texture the outside of the ring (see page 104) while it is on the mandrel. If you use this option to stretch your ring, file and sand the solder join first. Texturing with a steel hammer will stretch the metal. Make sure that you do not hit the mandrel with the metal hammer, as it will damage it.

FILING AND SANDING THE RINGS

1. On the round-wire ring, first tidy the solder join. Use the flat side of a hand file and move the file to follow the curve. If you do not follow the curve of the round wire, it will create a flat spot. Do not file the whole ring, as it will not need it. However, if you have put in any deep marks with the pliers, file those out now.

2. On the square-wire ring, first tidy the solder join. Use the flat side of a hand file to go over the solder join on the outside curve and the two flat sides of the ring. File until you cannot see the join, but do not make a thin area on the ring. Again, do not file the whole ring, as it will not need it. However, if you have put any deep marks in with the pliers, file those out now.

3. For the inside of both rings, use the curved side of a half-round hand file to file the solder area. Also file off any deep marks you may have created inside the ring.

4. For the round-wire ring, sand the filed areas. Use 500–600-grade abrasive paper on a stick (see page 55 for instructions on making a sanding stick) or wrapped around a hand file. Continue until all file marks are no longer visible.

5. For the square-wire ring, sand the three outside surfaces of the ring, including the file marks at the join. Use 500–600-grade abrasive paper on a stick. Sand in a different direction from the direction in which you filed previously so you can clearly see the file marks being removed.

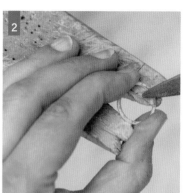

6. For the inside of both rings, use 500–600-grade abrasive paper wrapped around the curved side of a half-round hand file to sand out the file marks.

7. Repeat the sanding process above all over both rings with finer 1000–1200-grade abrasive paper. You may be able to remove the file marks using only the finer abrasive paper, skipping the 500–600-grade stage, but this will depend upon how deep the file marks are.

Now your rings are ready to be polished.

HANDY HINT

Each time you go down a grade, change direction so you can see the previous marks being removed.

HANDY HINT

At step 5, ensure that the abrasive paper is wrapped around something hard, otherwise the paper will round off the square corners.

POLISHING AND BURNISHING THE RINGS

Three methods of polishing are described on pages 54–57 in the Core Skills chapter. Choose from one of these methods:

1. Use a brass brush with water;

2. Rub with Brasso and polish off;

3. Use a barrel polisher.

Additionally, on the square-wire ring, you can rub the edges with a steel teaspoon or burnisher to highlight them (**4**).

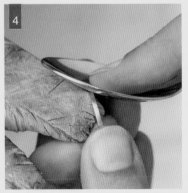

Your rings are complete.

PROGRESSION

Now that you have made these two rings, you have the skills you need to go on to make these pieces:

RINGS USING OTHER CROSS-SECTIONS OF WIRE

A variety of rings can be made from different cross-sections of wire.

BANGLES MADE FROM SILVER WIRE

An average-size bangle is made from wire 21cm (8¼in) or 22cm (8⅝in) long. The bangle should fit tightly over your knuckles where the fingers join the palm so that it is not too loose on the wrist.

When soldering a bangle, heat the whole bangle before focusing on the solder join.

CIRCLE PENDANT ON A CHAIN

Make a ring of any size with any cross-section of wire and thread a chain through it, or attach the chain with a lark's-head knot.

STACKING RINGS

Make a set of rings the same size and wear them together as stacking rings.

PROJECT 2:
TWO TWIST RINGS

The finished twist rings in square and round wire.

WHAT WE ARE MAKING

Two twisted wire rings:
one with round wire, one with square wire.

Making these rings will give you further practice in the core skills introduced in the first project. You will become more confident and understand these skills better, while learning additional techniques: twisting wire and a different way in which to size rings.

CORE SKILLS

Sawing; filing; soldering; forming; sanding; pickling; polishing.

NEW SKILLS

Annealing; twisting wire; another method of sizing rings.

YOU WILL NEED

BENCH, HAND AND FORMING TOOLS

- Piercing saw and saw blades
- Beeswax or candle wax
- Wire cutters (optional)
- Vice
- Hand vice (optional)
- Half-round pliers
- Hand drill
- Cup hook
- Ring sizers
- Ring mandrel
- Hide mallet
- Needle files
- Hand file

SOLDERING TOOLS

- Flux (such as borax) and dish or piece of slate
- Large soldering sheet
- Soldering block or charcoal block
- Paintbrush
- Torch for soldering
- Steel tweezers
- Tin snips
- Pickle in slow cooker
- Glass bowl with water and tongs (copper, brass or plastic)

FINISHING AND POLISHING TOOLS

- Abrasive paper
- Sanding stick (optional)
- Barrel polisher/brass brush or brass polish wadding (Brasso)

MISCELLANEOUS ITEMS

- Marker pen or pencil
- Masking tape (optional)
- Kitchen paper or cloth

MATERIALS

The lengths listed will make rings of most sizes except the very largest – i.e. above UK size V, US size 10¾. Add 10mm (⅜in) to the length of the wire for above-average ring sizes. These lengths allow for shrinkage due to the twisting of the wire.

- 21cm (8¼in) of 1.5mm round wire (15-gauge)
- 9cm (3½in) of 2mm square wire (12-gauge)
- Hard solder

PREPARING THE WIRE

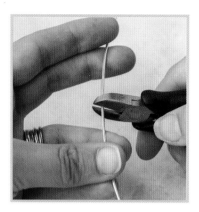

If the wires are not already at the required length, cut them with either a jeweller's saw or wire cutters – it is fine to use wire cutters because the ends of the wires do not need to be square for this project.

ANNEALING THE WIRE

If the wires that you are using are not brand new, they may need to be **annealed** (heated and softened). *To know if the wire needs annealing, hold the end of the wire in one hand, with at least 4cm (1⁹⁄₁₆in) sticking out. Spring it with your other hand. If it stays bent, it is soft and doesn't need annealing. If it springs back, it is hard and needs to be annealed.*

1. To anneal the round wire, first bend or curl it to fit on the soldering block. Move the flame around the wire to anneal it. Heat the silver until it is pink – this will be at a lower temperature than for soldering. Ensure that your lighting is not too bright as you might not see the wire turn pink. However, you will see the wire relax as it anneals and it will drop down onto the soldering block.

2. To anneal the square wire, lay it in a line pointing away from you on the soldering block so that you can heat it efficiently. Start at the end nearest to you and then move the flame along the wire up to the other end. If you start at the end furthest away from you and move towards you, the end you have already heated is still in the flame and you run the risk of melting the wire.

3. Pickle and rinse the wire to clean it.

TWISTING THE WIRE

THE ROUND WIRE

Fold the wire in half into a 'U' shape. Place the ends of the wire close together and one above the other (not overlapped). Clamp them into the side of the vice tightly so the wire cannot slip out – see photograph 1, below. Put only approximately 1cm (⅜in) into the vice, otherwise you are wasting wire. Check that they are clamped securely by pulling on each wire.

1. Put a cup hook into the chuck of the hand drill. Hook it through the loop of the round wire. Keep the wire taut and parallel to the floor.

2. Turn the handle of the drill in one direction until you get an even twist along the wire.

3. Keep going until you are happy with the tightness of the twist.

4. Unhook the cup hook and remove the wire from the vice.

HANDY HINT

Put masking tape around the screw thread of the cup hook before putting it in the hand drill – this will help the drill to grip the hook more securely.

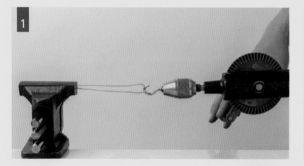

Above, the optimum tightness of your twisted round wire.

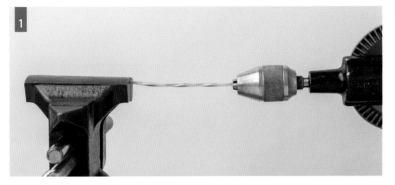

THE SQUARE WIRE

1. Place just under 1cm (⅜in) of the end of the wire in the vice. Secure the other end directly in the chuck of the hand drill. Keep the wire taut and parallel to the floor.

2. Turn the handle of the drill in one direction until you get an even twist along the wire.

3. Keep going until you are happy with the tightness of the twist.

4. Remove the wire from the vice and the chuck of the hand drill.

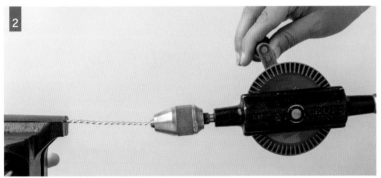

HANDY HINT

When a twist is uneven, it needs to be annealed where it is looser. The twist is looser where the wire is harder, meaning it has more resistance at that spot. Anneal the sections where the twist is looser, making these sections softer than the other parts. Put the wire back in the vice in the same way and repeat the steps above to even out the twist.

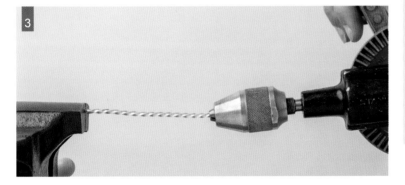

GOOD AD-VICE

If you are a tool fiend, like we authors are, this is a good time to buy yourself a hand vice (right). A hand vice grips the wire really well and will allow you to twist thicker wire that may not fit in the chuck of the hand drill.

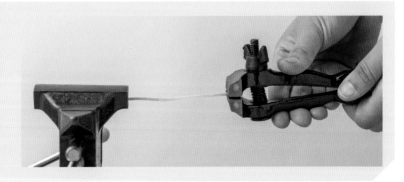

SIZING AND FORMING THE RINGS

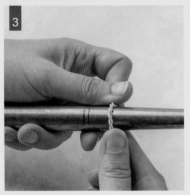

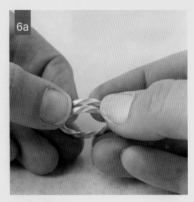

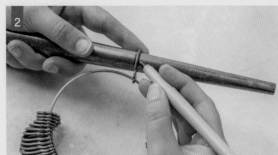

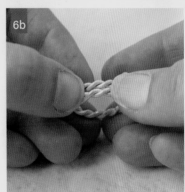

1. Twisting will harden the wire, so you will need to anneal both wires to make them malleable – see page 78.

2. Measure your finger size with ring sizers. Put the appropriate ring sizer on the mandrel and mark its position on the mandrel with a marker pen or pencil.

3. With your fingers, bend one of the wires around the mandrel so that it is smaller than the size that you have marked. You will need to start smaller as the wire will spring back a bit. If the wire is difficult to bend by hand, use a mallet to form the wire around the mandrel.

4. The ends of the wire will be difficult to push down by hand – use a mallet to tap these down on the mandrel into a round shape in order to get an overlap.

5. Push the ring down the mandrel until it matches the size at the mark you have made at step 2.

6. Adjust the ring overlap so that the twist matches up. When adjusting, make the ring a bit smaller than required, not bigger, so that the ring can be stretched if needed. Making the ring too big may result in you having to cut out a piece: **a)** shows the square-wire ring being adjusted; **b)** shows the round wire being adjusted.

7. Cut through the ring overlap. When you are nearly at the end of the cut, remember to cut through the wood at the same time, to avoid creating a burr.

8. Refer to page 67 to check that the join is good, and learn how to correct it if it is not.

HANDY HINT

If you need to file the ends of the round-wire twist to get a good solder join, it is helpful to run some solder on both ends of the twist before filing, as otherwise the round wire can unfurl as you file it.

SOLDERING THE RINGS

1. For the twisted square-wire ring, use hard solder to solder the join as you did in project 1 (see pages 68–69).

2. For the twisted round wire, the join is effectively two joins that are close together. Place two pieces of hard solder, one on each join, and solder at the same time. The solder may flow around the twist as well as on the join area. This is not a problem. Once you see the solder flowing, remember not to linger with the flame, as capillary action will pull the solder away from the join and around the twist.

3. Pickle and rinse to clean.

Two options for soldering the round ring.

Left, the join only has been soldered; right, the solder has been flowed all the way round the twist. If you want the solder to flow all the way around, you will need to use more solder.

FORMING AND FILING THE RINGS

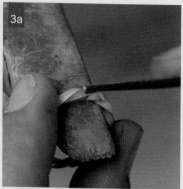

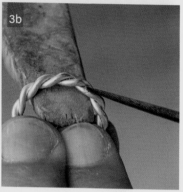

1. For both rings, put each on the mandrel in turn and use a mallet to make them round.

Note: If the ring is too small, stretch it by hammering it with a mallet down the mandrel. If the ring is too big, you will need to cut out a piece and resolder. See Adjusting the Size of a Ring, page 71.

2. Check that the rings lie flat on a hard surface. If not, put a few layers of kitchen paper or a cloth on the steel block and tap the ring with the mallet. The padding is to prevent making flats (i.e. damaging the edges of the wire) on your rings. Remember to keep your fingers upright.

3. To tidy the rings at the solder join, use needle files. They are small enough to get into the twist.

a) For the square-wire ring, use a half-round or round needle file and file following the twist.

b) For the round-wire ring, use a needle file with a flat face that allows you to follow the twist. A triangular file is good for getting into the twist of the round-wire ring. Remember to file following the curve of the ring to avoid making the ring thinner where the solder join is.

HANDY HINT

If the twists do not quite line up, they can be disguised with filing so that the ends look like they do join up. You can do this by filing the ends that are out of line so that they blend into each other. Do not overdo this, however, as it can create a thin area on the ring.

SANDING AND POLISHING THE RINGS

1. Sand both rings to remove the file marks and any other marks using the finer 1000–1200-grade abrasive paper. It helps if you fold the abrasive paper to make it stiffer in order to get into the detail and take out the file marks. You can also wrap abrasive paper around the needle files you have already used to file the rings.

2. If the edges of the square-wire ring feel sharp, soften them by wrapping a piece of abrasive paper around the ring and sanding off the sharpness.

Both rings are ready for polishing. You can use either a barrel polisher or a brass brush.

You have now made four rings, using two different methods to size the rings.

PROGRESSION

Now that you have made these two rings, you have the skills to try other types of twisting. Experiment with different twists and wires, and see what looks good.

A square-wire bangle made by twisting alternating sections of the wire to the right and to the left:

1. Twist sections of square wire to the right and then to the left in turn.

2. Mark off equal sections along the wire.

3. Place the wire in the vice at the first mark and place a hand vice at the next mark. Twist that section, counting the number of turns.

4. Remove the wire from the vice.

5. Wrap kitchen paper or leather around the twisted part.

6. Place back in the vice where the twist finishes.

7. Put the hand vice at the next mark and twist the same number of turns in the opposite direction. Continue, repeating this pattern, down the whole length of the wire.

You can also do the above, twisting sections of the wire, but alternate with sections that you do not twist at all, as shown on the square-wire bangle on the right.

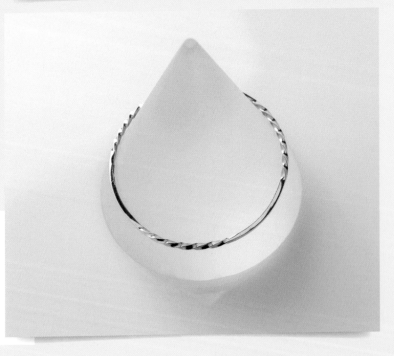

GENERAL NOTE

If you are using thick wire, start with a longer piece of wire, because it will shrink in the twisting process. Remember, the thicker the wire, the harder it is to twist.

Two round wires of different metals can be twisted together, such as copper and silver. If they are the same thickness, you can clamp both ends in the table vice and the hand vice to twist. If they are not the same thickness, solder both ends together side by side, then twist by holding both ends in vices (see Good Ad-vice, page 80).

A round-wire twist can be hammered flatter on a ring mandrel with a steel hammer after twisting.

Use other cross-sections of wire, such as rectangular or oval, for different twist effects using the method for the square wire in the project.

MAKING JUMP RINGS

Jump rings are circles made of wire that are used for a wide variety of functions, including connecting jewellery parts such as clasps to chains, charms to bracelets, and much more.

It is much easier to make jump rings from a single, long length of wire, so form the jump rings first and then saw them, whether you are making one or multiples. It is useful to practise making jump rings with copper wire.

HOW TO MAKE ONE JUMP RING

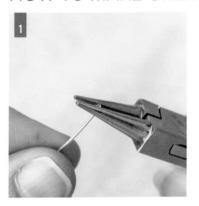

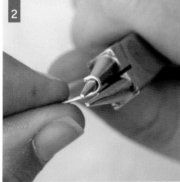

1. Use round-nose pliers (or round-and-flat-nose pliers) to form the jump ring. Where you wind the wire on the pliers dictates what size the jump ring will be. Grip the end of the wire in the round-nose pliers and use your thumb to guide the wire around the nose of the pliers while twisting the pliers.

2. One twist of your wrist will make about half the jump ring. To continue, release the pliers, spin them around in the curve and grip the wire again and twist the pliers again to form the rest of the jump ring. If you are using round-nose pliers, do not grip the wire too hard or the nose of the pliers will mark the outside of the jump ring. Using the round-and-flat-nose pliers avoids marking the outside of the wire.

3. Once you have a circle, keep going until there is an overlap. Make sure that the end of the wire that is already formed is winding towards the narrower end of the pliers, to keep the jump ring round. When bending any wire, often the end does not form into a neat curve, hence the need to go beyond the end of the wire to make an overlap. Otherwise the jump ring will not be round. The overlap should not have a gap between the wires, to make sawing easier.

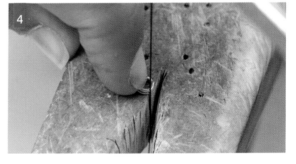

4. To saw, grip the length of wire that the jump ring is on with your hand and hold down the jump ring on the benchpeg with your index finger. Position the jump ring so that when you cut through the overlap the saw will hit wood before hitting the other side of the jump ring.

5. Saw through the overlap. You may find it easier to position the jump ring so that you cut into the wood at the same time. This avoids making a burr.

HOW TO MAKE A FEW JUMP RINGS

This is the method you will use to make the jump rings in the Stone-chip Earrings project (see pages 92–103).

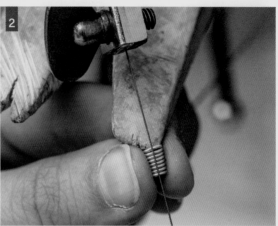

HOLDING THE COIL

If you are struggling to hold the coil, you can use masking tape to make it easier. Cut a small piece of masking tape and stick it onto the front of the coil where you will cut. Pull the two sides of the tape to the back and stick them together to make a tab. This creates a handle to help you hold the coil while you are sawing it on the peg.

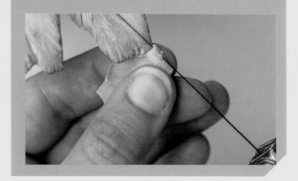

1. Start in the same way as for making one jump ring with the round-nose pliers. Don't stop when one jump ring has been formed: continue to form a coil. Ensure that you are winding at the same spot on the pliers so the jump rings are the same size. The end of the coil must wind off the pliers towards the narrower end. Wind the coil so that there are no gaps between the jump rings.

2. To saw the jump rings, hold the coil tightly between your index finger and thumb and the rest of the wire is secured in your hand. This wire helps you to hold the coil as you are cutting: do not cut off the excess until you have cut through the last jump ring. Your hand should be under the coil. Position the end of the coil against the end of the benchpeg so that the corner of the benchpeg is nestled inside the end of the coil.

3. Push the coil towards the benchpeg as you are cutting to keep it steady.

4. Saw through the top of the coil where it is resting on the benchpeg. As you cut through the jump rings, let them fall off the coil. You can arrange your apron across your knees to catch them. Sawing these jump rings is not easy, but it will become easier the more you do it. Sawing is better than using wire cutters, as you end up with good, clean solder joins.

HOW TO MAKE MANY JUMP RINGS

Choose a **former** for the size of jump ring that you want to make. You could use a knitting needle as we have done; alternatively, you could use a steel rod, a skewer, wooden dowelling, a pen, the handle of a doming punch or a nail – anything that is cylindrical and sturdy enough to wrap wire around. You will need to ensure that you can remove the coil easily, once formed.

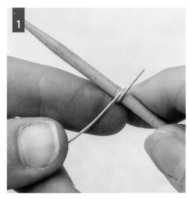

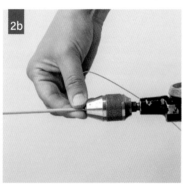

1. One way to wind the wire around the former is by hand. Don't start at the very end of the wire or you will not be able to stop it spinning around the former. Form a coil around the former, leaving no gaps between the jump rings.

2. Another way to form the coil is by placing the former in the chuck of a hand drill.

a) Put the drill in a table vice so that the winding handle is above the vice, so it can turn freely.

b) Push the end of the wire in between the jaws that are holding the former and bend the wire towards you so that it bends at 90 degrees away from the former. Hold the wire between your index finger and thumb against the former in order to guide it.

c) Turn the drill and the wire will coil around the former as it turns in the drill.

d) When the coil is long enough, remove the coil and the former together from the drill.

With either method, when you have made a coil long enough for what you need, go back to the end of the wire and bend it around the former to complete the coil. You will need to use chain-nose or flat-nose pliers to do this.

To saw, use the same method as for a few jump rings (see page 89). You can keep the coil on the former to keep it steady, making sure that you pull the former out of the section of the coil you are cutting.

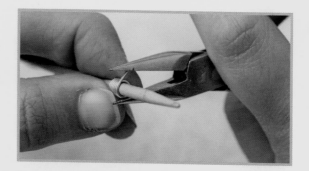

TROUBLESHOOTING

PROBLEM

Once the jump ring has been sawn, there is a burr which stops it closing.

A burr in the join of the jump ring.

SOLUTIONS

• If the burr is very thin, it can be knocked off when the jump ring is closed with two pairs of flat-nose pliers (or a pair of flat-nose pliers and a pair of chain-nose pliers).

• If the burr is bigger, use a flat file to file it off. You might need to twist the ring open to avoid filing any other part of the jump ring.

PROBLEM

When sawing, the jump rings keep bending and breaking the saw blade.

SOLUTIONS

• Cut the jump rings with wire cutters and file the ends flat if the jump rings are big enough.

• Use a finer saw blade, such as a 6/0.

STONE-CHIP EARRINGS

The finished stone-chip earrings.

WHAT WE ARE MAKING

A pair of hook earrings with stone chips.

Making these earrings will introduce you to many more techniques, including how to secure components when soldering is not possible.

These earrings have an added dimension of colour.

CORE SKILLS

Sawing; forming; soldering; pickling; polishing.

NEW SKILLS

Making jump rings, headpins, earring hooks and wire wrap loops; pick soldering; soldering links of a chain.

YOU WILL NEED

BENCH, HAND AND FORMING TOOLS

- Piercing saw and saw blades
- Beeswax or candle wax
- Needle files
- Wire cutters
- Round-nose pliers
- Flat-nose pliers
- Chain-nose pliers
- Round-and-flat-nose pliers (optional)
- Ruler
- Scriber

SOLDERING TOOLS

- Flux (such as borax) and dish or piece of slate
- Large soldering sheet
- Soldering block or charcoal block
- Paintbrush
- Torch for soldering
- Steel tweezers
- Reverse-action tweezers
- Tin snips
- Pickle in slow cooker
- Glass bowl with water and tongs (copper, brass or plastic)

FINISHING AND POLISHING TOOLS

- Abrasive paper
- Sanding stick (optional)
- Barrel polisher, brass brush or polishing cloth

MISCELLANEOUS ITEMS

- Masking tape (optional)
- Pen (for forming the earring hooks – see page 99)

MATERIALS

- Six semi-precious stone chips
- A length of 0.8mm (20-gauge) round wire no shorter than 21cm (8¼in) for the earring hooks and jump rings
- For the headpins, either approximately 30cm (12in) of 0.8mm (20-gauge) round wire or, if the holes in the stone chips are too small for that wire, 30cm (12in) of 0.6 (22-gauge) round wire
- Hard and easy solder
- A length of copper wire (optional, for practising making wire wrap loops – see page 100)

MAKING THE JUMP RINGS

Here, we are going to make two small chains out of jump rings, which will be connected to the earring hooks. Start by making six jump rings with 0.8mm (20-gauge) round wire, using the method shown on page 89, How to Make a Few Jump Rings. Make these jump rings as big as you can by using the widest part of the round nose pliers. Close four of the jump rings following the instructions below.

CLOSING THE JUMP RINGS

1. With a pair of flat-nose pliers in one hand and a pair of chain-nose pliers in the other, grip the jump ring on either side of the opening.

2. Twist the ends of the jump ring in the pliers until they line up. To get the ends completely in line, you might have to twist the ends of the jump ring beyond lining up, as the metal will spring back a bit.

3. Make sure that there is no gap so the join is ready for soldering. If there is a gap, twist open the jump ring and push the ends past each other a little to create tension when you twist back to close.

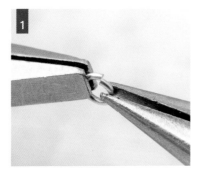

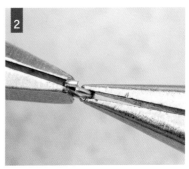

HANDY HINTS

Do not open a jump ring by pulling the ends away from each other. This will distort the shape. Instead, twist open or closed sideways using flat-nose pliers or chain-nose pliers on either side of the join.

SOLDERING THE JUMP RINGS

Here, you will practise four different ways of placing the solder on a join to build up your experience and understanding of soldering. Place the four closed jump rings from left to right with some space between them on the soldering block. Have the soldering joins away from you – this will reduce the risk of melting each jump ring.

For three of the jump rings, try these ways of placing the hard solder:

1. On the first jump ring, place a pallion of solder on top of the join.

2. On the second jump ring, place a pallion of solder on the soldering block against the join.

3. On the third jump ring, place a pallion of solder under the join by putting it on the soldering block and placing the jump ring join on top of it.

4. Solder the jump rings one at a time. To solder the first two jump rings, the flux needs to be dried by dipping the flame on and off. Reposition the solder if it moves. Once the flux is dry, flow the solder. Do not pickle.

5. To solder the third ring, you do not need to dip the flame on and off because the solder is held under the jump ring, so cannot move (as long as your flux is not too thick). When you heat this jump ring, you will see the ring drop down when the solder flows. Do not pickle.

6. To solder the fourth jump ring, use the pick soldering method shown overleaf.

PICK SOLDERING

Pick soldering is a different method of placing the solder. It involves using a metal pick to pick up the solder and place it on the join. The pick is usually made of steel but can be titanium. You can buy a pick or use tweezers or an unfurled paperclip with the end filed to a taper.

1. Place a pallion of solder on the soldering block, ready to use.

2. Holding a pick behind the pallion, melt the pallion into a ball. In one action, remove the flame and scoop up the ball of solder with the point of the pick while it is still hot. The ball of solder should stick to the end of the pick. If it does not stick, heat the solder and try again.

3. Without heat, practise the motion of putting the solder on the join with the pick so that you know you can reach the join with a steady hand. It is important that your hand is steady so that you have control when placing the solder. Rest your arm or hand against the table or a safe part of the soldering area to keep it steady.

4. Now, with the torch lit, hold the pick ready to place the solder on to the join. Make sure that the flame is facing forwards and not at the hand holding the pick.

5. Heat the join. When the flux has dried, keep the flame on it and gently introduce the solder on the join. Don't push the solder against the join, as it will break away from the pick and roll across the soldering block.

6. When the solder has transferred from the pick to the join, move the pick away.

7. Continue to heat until the solder flows.

TROUBLESHOOTING

1. When melting the pallion of solder into a ball, it can roll off the soldering block. Place the pick behind the pallion to stop this happening.

2. If solder comes away from the pick while you are trying to put it on the join, it is because the pick is scraping against the piece. You need to present the piece of solder to the join, not try to scrape it on.

3. If the solder does not transfer from the pick to the join, it is because the join is not hot enough.

4. If the pick is getting too hot to hold, the join is not yet hot enough so you are introducing the solder too early. You may also be heating the pick instead of the join.

HANDY HINT

If you are using tweezers, you can pick up the pallion and melt it into a ball while holding it within the ends of the tweezers. When you release the pressure, the ball will stick to one inner side of the tweezers.

However, because the solder is on the inside rather than the outside of the tweezers, you will not be able to reach some types of joins because the tweezers will be in the way.

SOLDERING THE JUMP RINGS TO MAKE TWO CHAINS

1. After quenching the jump rings, place two of them on one of the remaining open jump rings and then close it (see page 94). Repeat with the other two soldered jump rings so that you have two three-ring chains.

2. To set up for soldering, grip the unsoldered jump ring with the tip of the reverse-action tweezers away from the join. Do not cover the join with the tweezers.

3. Place the reverse-action tweezers on the soldering block with the soldered jump rings hanging off the block so that the unsoldered join is uppermost.

4. Flux the join.

5. Place a small pallion of easy solder on the join. There is more chance the pallion will stay on the join if the flux is still wet, as it sticks to the flux. If the flux is dry, the pallion may fall off.

6. Hold the flame horizontally to avoid heating the two lower jump rings. Heat above the join until the flux is dried. If you aim the flame directly at the join before drying the flux, it will blow the solder off. When the

flux has dried, the pallion of solder will stick to the join. Now bring down the flame onto the join until the solder flows. The reverse-action tweezers act as a heat sink (i.e. they conduct some of the heat away from the jump ring), so be aware that you need to heat the side the tweezers are on a bit more to get an even heat across the join.

7. Set up the second three-ring chain on the soldering block in the same way, ready for pick soldering.

8. To practise pick soldering again, we're going to solder the other jump ring using this method. Place a pallion of easy solder on the soldering block. Pick up the solder with your pick as described on the previous page. Practise positioning your hand first so that the pick with the solder on is in the right place. Now heat the jump ring join and transfer the solder from the pick to the jump ring join when it is hot enough. Flow the solder.

9. Pickle both the three-ring chains.

MAKING THE HEADPINS AND THE EARRING HOOKS

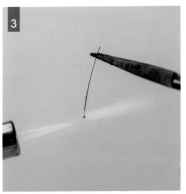

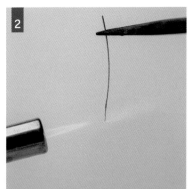

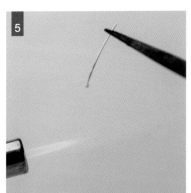

Cut two 60mm (2⅜in) lengths of 0.8mm (20-gauge) round wire with wire cutters.

Cut six 50mm (2in) lengths of the thickness of wire that goes through the holes in your stone chips. This will be either 0.8mm (20-gauge) or 0.6mm (22-gauge) round wire.

1. Place all the pieces of wire on the soldering block with the ends hanging over the edge. This enables you to pick them up one at a time with the reverse-action tweezers while the torch is lit.

2. Light the torch and pick up one wire with the reverse-action tweezers. Hold the wire so that it is pointing downwards with the tweezers towards the top. Place the bottom end of the wire in the flame at the point of the light blue cone.

3. You will know that the wire is in the flame when you see an orange glow behind the wire.

4. The end of the wire will melt up into a ball. If you have the flame too high up the wire, you will burn through the wire and the bottom will fall off. As the ball forms, follow it with the flame.

5. If the ball gets too big, gravity will make it fall off. The balls only need to be approximately 1–1.5mm (1/32–1/16in) wide, so that the stone chip will not come off the wire.

6. Put the wire back on the soldering block and pick up another one.

7. Repeat steps 1–6 with the remaining seven pieces of wire, then pickle all the wires.

HANDY HINT

If you are using this technique with thicker wire, it is a good idea to first flux the ends you are balling up.

FORMING THE EARRING HOOKS

The former for making the earring hooks can be a cylindrical item such as a pen, pencil or steel rod – we are using a pen in the photographs shown below. Consider the diameter of your former carefully so that you get the shape you want for your earring hooks.

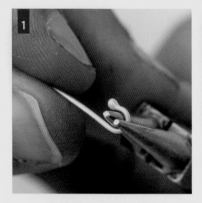
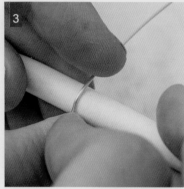
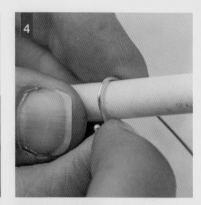

1. Use the two longer pieces of wire to make the earring hooks. Grip one end behind the ball with the middle of the round-nose pliers. Bend the wire to make a 'U' shape.

2. Hold the 'U' between your index finger and thumb.

3. Hold your finger and thumb against the side of the former.

4. Push the long end of the wire over and down around the former.

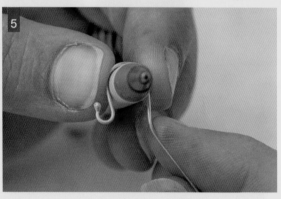

5. With the wire on the former, grip the end of the wire (not the ball end) with your finger and thumb and put a slight curve on it, bending away from the ball side. You can go on to experiment with earring hook shapes to find the one you like best.

6. Trim the two earring hooks to the same length.

7. To take the sharp edges off the ends of the wire where they were cut, dome the ends by filing with a flat file.

IMPORTANT NOTE

If the stone chips you are using are made of softer stone like turquoise, or anything below 7 on the Mohs scale of hardness (see pages 190–191), you will not be able to put them in a barrel polisher as it will damage them. The same goes if you are unsure your stone chips will survive the barreller. Instead, put the chains on the earring hooks and barrel-polish these with all the end pins at this stage. Once polished, add the stone chips to the end pins then follow the instructions on the following page for making wire wrap loops.

MAKING THE WIRE WRAP LOOPS AND ADDING STONE CHIPS

You may wish to practise with a length of copper wire first to get used to making the loops.

> **HANDY HINT**
>
> *Remember to put the chain onto the wire wrap loop after you have formed the first loop and before you start wrapping, or it will be too late. See photograph 6, below.*

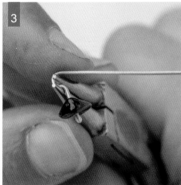
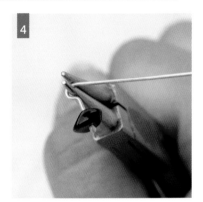

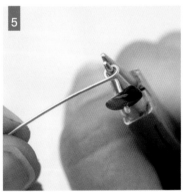
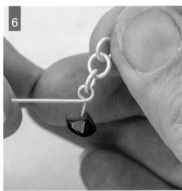

1. Put one stone chip on each of the remaining balled-up wires (called headpins). If you have a choice of stone chips, choose ones where the hole is in the middle and not too close to the edge. Grip the wire where it comes out of the stone chip with chain-nose pliers. Bend the wire with your fingers against the pliers to make a right angle.

2. Remove the chain-nose pliers. Grip the wire with round-nose pliers in the corner of the bend with one nose above the other. The further up the pliers you hold the wire, the bigger the loop will be.

3. Bend the wire over the top of the round-nose pliers to start the loop.

4. Move the pliers up so that the bottom nose is now inside the loop and the other nose is above it.

5. Bend the wire under the bottom nose to complete the circle. This end of the wire should be at right angles to the part coming out of the stone chip. Repeat with all the remaining wires.

6. Thread one of the wire loops with a stone chip through the last link of one of the three link chains.

7. Grip the wire loop that has the stone chip on it with chain-nose pliers.

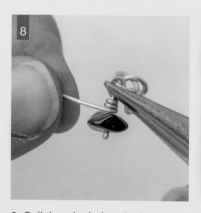

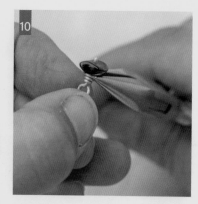

8. Coil the wire below the loop, working down towards the stone chip. Make sure that you keep the wire you are wrapping at right angles to the wire you are winding it around, otherwise the coil will be too open. If the wire is very short and you cannot coil it with your fingers, use chain-nose pliers to bend the wire around to form the coil.

9. If there is still wire left when the coil has reached the stone chip, cut off the excess with wire cutters.

10. In both cases, the very end of the wire will need to be tucked in to finish the coil using chain-nose pliers. Hold the chain-nose pliers on either side of the middle wire and squeeze the wire down to form the coil.

11. Add two more to the same link of that chain. Then there will be three wire loops with stone chips on the chain. Repeat for the second chain.

ADDING THE EARRING HOOKS AND FINISHING THE EARRINGS

1. Loop the other end of one of the chains (the end without the stone chips) over the ball of the earring hook.

2. Push the ball back against the wire, closing the loop so that the chain cannot come off the earring hook.

3. Repeat for the other earring hook.

4. If you haven't polished the component parts previously (see Important Note, page 99), then barrel-polish now. You can also polish with a polishing cloth or brass brush.

PROGRESSION

Make these earrings with beads instead of stone chips.

Make the earrings longer by adding more jump rings to the chain.

Make the earrings with stone chips or beads in each link, instead of all at the bottom.

Make bigger links with thicker wire to make larger chains, using larger formers.

Make a ring with a half loop on it to add wire wrapped stone chips or beads to decorate.

Onto a large jump ring that has been soldered closed, hang stone chips with wire wrap loops and spirals made of wire to form a decorative cluster. You can then thread a chain through the large jump ring.

TEXTURING METAL

The textured pendant project (see pages 112–121) and the brooch project (see pages 146–161) both invite you to texture silver sheet. Before you commit any texture to silver sheet, practise some or all of the texturing methods on the following pages on a piece of copper sheet that is approximately 35 × 35mm (1⅜ × 1⅜in) and 0.7mm (21-gauge) thick.

In the examples that follow, one photograph shows how to apply all the textures on a single piece of copper sheet. The additional photographs show how the texture might look on a whole piece of sheet.

1. HAMMERING

Hammering is a way of creating many different textures to add interest and new dimensions to pieces of jewellery. Remember to hold hammers at the end of the handle so the head of the hammer bounces on the metal, rather than trying to punch the metal. If you are doing a lot of hammering, use ear plugs or ear defenders.

a) A ball pein hammer is a steel hammer with a flat face and a domed face. Place the metal on a steel block and hammer it with the domed face of the hammer. This creates a dimpled surface.

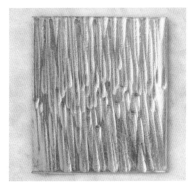

b) A cross pein hammer is a steel hammer with a flat face at one end and a straight face at the other. Place the metal on a steel block and hammer it with the straight face to create a lined surface.

HANDY HINT

If you are not getting a complete line indentation, then you may be holding the hammer at an angle. Adjust how you are hammering so the hammer head is flat to the metal. (Or maybe you like the marks as they are!)

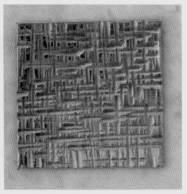

c) Using the cross pein hammer, work over the surface again in another direction to create a different effect.

2. PUNCHING

When using punches, place the metal on a steel block. Hit the punch once with a hammer or mallet. Try to avoid hitting the punch multiple times to prevent making a double print.

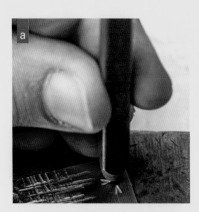

a) Sets of steel letter and number punches are available in a wide variety of sizes and fonts. These can be used for words or patterns. It takes a lot of practice to stamp words evenly. Here, the letter punches are being used to make patterns. Play with the placement of different letters and numbers to see what patterns you can make. For example, use the 'V' to make a decorative edge (top right); use the 'C' (bottom left) or 'O' (bottom right) to make fish-scale patterns.

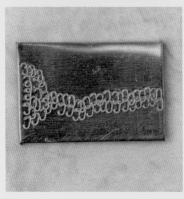

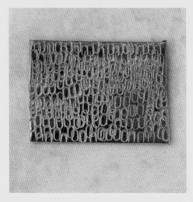

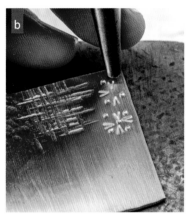

b) Centre punches are steel punches with a blunt point at the end, normally used to make an indentation to place a drill bit. You can make a centre punch by filing a blunt point on a large nail. These punches make dots that can be used to make myriad patterns either by themselves or in combination with other texturing methods. In the photograph on the far left, dots are combined with the stamped letter 'V'.

c) Doming punches are used with the doming block to make discs into domes. For texturing, these can be stamped around the edge of the metal to give a piecrust edge to the sheet. Be very careful not to let the doming punch come into direct contact with the steel block, or you will create a flat on the domed surface. You can also use a ball pein hammer as a punch by placing it on the edge of the sheet and hitting the flat face with a mallet.

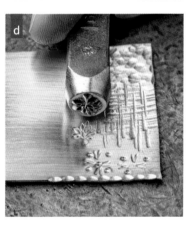

d) Decorative punches are steel punches manufactured with a wide variety of designs (on the far left, a tree). These can be used alone or in combination with other texturing methods. These designs are usually larger and need a heavier blow from a mallet or heavy hammer to create a full indentation. It helps to stand while hammering, to deliver a heavy blow.

Don't expect to get a perfect imprint when you start – do practise. The punch needs to be vertical to the surface of the metal or you will imprint only one side of the design. To get a good imprint, it will help to make sure that your metal is **annealed** before you punch it (see page 78).

e) You can make your own punches by cutting the point off a large round nail or a steel rod (using a hacksaw). Make the end flat with a file and then use a file or a saw to create lines or shapes in the end. The same can be done with the faces of old hammer heads.

If you do use a file, ensure that it is used only to file steel.

3. WIRE

Wire can be used to make shapes that are hammered into the metal. Create the wire shape using your hands, pliers or formers. Place the shape on the metal and tape it down with masking tape. Hammer using the flat face of a steel hammer. Make sure that the hammer head is flat to the surface of the metal or you will mark the metal with the edge of the hammer. The wire leaves a print of the wire shape you have made in the sheet.

Do not overlap the wire at any point because it will not imprint well at that point.

Any metal wire (such as copper, brass or silver) can be used for this method. Binding wire is a mild steel wire usually used to hold things in place for soldering. It is harder than the metal you are stamping, so will leave a strong mark. You cannot reuse the same piece of wire to make a repeat pattern because this method flattens the wire. Note that the wire may leave a mark on the face of the hammer, so it is best to use a general-purpose hammer and not one with a polished face.

This tulip is made up of different pieces of wire that were stuck to the masking tape before it was attached to the metal. The pieces of wire were placed on the masking tape using tweezers.

4. USING A SCORPER

A scorper is a steel cutting tool used for carving or removing metal. The one being used here for texturing has a flat cutting edge that looks like a chisel.

To texture with the flat scorper, walk it across the metal on the corners of its cutting edge, making a zigzag pattern across the surface. It can be used in straight lines or curves. Going over the pattern again in a different direction results in an even patterned surface so you cannot see the direction in which you have used the scorper.

You can also make a basket-weave effect. To do this, mark out the width of the flat of the scorper by rocking it in a line across the metal so that the corners leave a dot. Then, in between every alternate dot, walk the scorper from one dot to the next in one direction, to make a small square of texture. Go back and, in the alternate blank spaces, walk the scorper 90 degrees to the marks already made. You will have a line of alternating squares in a zigzag pattern. Go back to the beginning and do another line alongside this one, making sure that you work at 90 degrees from the adjacent square. Keep building up this pattern until you have the amount of basket-weave pattern that you want.

Left to right: straight lines next to each other, basket-weave, straight line, curved line.

5. SATINIZING

Satinizing creates a scratch-surface finish. Note that this cannot be used to disguise bad surfaces: you will still need to file and sand your piece up to polishing stage before satinizing, otherwise the bad surface will show through.

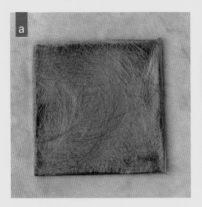
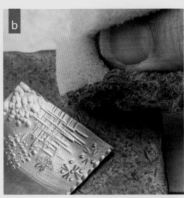
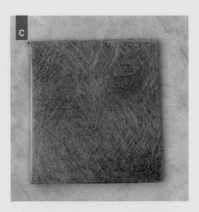

a) Coarse abrasive paper can be used in a circular motion over the area to be satinized. It is best to use it in a circular motion, not backwards and forwards, because you can see where you stop when you go backwards and forwards.

b) A scouring pad (or the rough side of a dishwashing sponge) can be used in a circular motion, or a back-and-forth motion, as it doesn't leave directional lines.

c) Abrasive blocks are available in different grades from fine to very coarse. The very coarse block can be used to give a satinized finish.

HANDY HINT

On some items of jewellery, especially rings, the satinized surface can become more polished as it is worn. To resatinize the surface, repeat any of the methods above.

6. BURNISHING

Burnishing uses polished steel to make a shiny surface on edges or high spots (though not on the flat part of a sheet). A burnisher is a tool made with polished steel in a wooden handle. If you don't have a burnisher, you can use the back of the bowl of a steel teaspoon or the back of a pair of round-nose pliers.

Use by firmly rubbing along the area you wish to put a polish on. The compression of the metal when you push the burnisher over it is what creates a polished surface. This works well in combination with satinized pieces as it puts a shiny frame around the edge of the piece.

7. USING A PIERCING SAW

A piercing saw can be used to saw decorative lines. Usually this method is used on wire, but it can also be used on the edges of metal sheet.

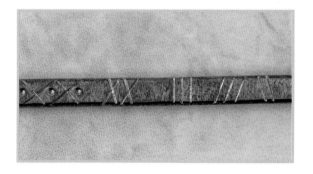

Use the saw to cut lines: these could be parallel lines, 'V' shapes, 'X' shapes, or whatever you can come up with. The lines can be combined with any other texturing methods, and then oxidized (see opposite page) to make them stand out.

Note that when cutting lines on a wire with a curved profile you should not saw straight downwards or the cut will go too deep. Instead, saw following the contour of the wire.

8. OXIDIZING

Oxidizing is the process of blackening the metal with chemicals. It is used to blacken details, such as texture, to give a contrast or to give an antique look. Changing the colour of the surface of any metal is called **patination** and the resulting colour is called a **patina**. Most chemicals used to oxidize silver have a strong, unpleasant sulphuric (egg) smell. Below are three different ways to oxidize silver.

SAFETY NOTE

Wear gloves and goggles when handling any oxidizing chemicals.

HANDY HINT

When polishing patinated pieces, don't use a brass brush as it will go in the recesses and remove the patina. Pieces with patination can be barrel-polished.

You can use plastic tweezers to dip items into liver of sulphur solution or Platinol, as the solutions do not affect them. If you use metal tweezers, be sure to rinse them after use.

a) Boil an egg, peel it and cut it into pieces. Place your piece and the egg (not touching) into a plastic food box. Put the lid on and leave it until the metal goes black (this could happen overnight). This method blackens the whole piece. To show the texture, sand over the surface, leaving the black in the recesses.

b) Liver of sulphur can be used to create yellow, brown, blueish and blackish patinas. You can buy it in pre-mixed liquid form, solid lumps (which need to be dissolved in water when you want to use it) or gel. Store out of sunlight. To use, prepare one container of warm water and one container of cold water. Pour a bit of the liver of sulphur into the warm water. Dip your piece into the solution and take it out with tweezers to look at the colour. Keep it over the container so the drips are caught in it. If you like the colour, rinse in the cold water to stop the reaction from continuing. You can repeat this process until you are happy with the

colour. Again, this patinates the whole piece, so you can leave it all over or sand over the surface to brighten the high spots. Once you have gone past a colour, you cannot go back to it unless you sand off the patina and start again.

c) Platinol (an oxidizing solution) is a commercially available solution which makes a strong dark patina. Use a synthetic paintbrush (as natural bristles will dissolve) to apply the Platinol where you want the patina (above, left). For fine areas, you can apply it with a cocktail stick or toothpick. Rinse in water once applied. This is the only method where you do not have to darken the whole piece. Areas can be picked out without affecting the rest of the piece. However, you can immerse the whole piece and sand back the surface as with the other methods. Platinol works best when used in one go, as if you try to layer it up the surface can start flaking.

PROJECT 4:
TEXTURED PENDANT

The finished textured pendant.

WHAT WE ARE MAKING

A textured, domed pendant.

Texturing can be used to enhance your jewellery and give it interest. Here, we use it to make a picture. You can follow these instructions or choose your own texture from the previous pages.

CORE SKILLS

Sawing; filing; forming; soldering; sanding; pickling; polishing.

NEW SKILLS

Marking out onto sheet; cutting a circle out of silver sheet; texturing metal; doming; drilling; making a bail.

YOU WILL NEED

BENCH, HAND AND FORMING TOOLS

- Dividers or circle template
- Scriber
- Piercing saw and saw blades
- Beeswax or candle wax
- Hand file
- Ball pein hammer
- Doming block
- Doming punches
- Hand drill
- Drill bit, between 1.0–1.6mm (¹⁄₁₆in)
- Centre punch
- Chain-nose pliers
- Flat-nose pliers
- Ruler
- Hide mallet
- Binding wire (optional)
- Cross pein hammer (optional, for other textures)
- Punches (optional, for other textures)
- Scorper (optional, for other textures)
- Burnisher (optional, for other textures)
- Liver of sulphur or Platinol (optional, for other textures)

SOLDERING TOOLS

- Flux (such as borax) and dish or piece of slate
- Large soldering sheet
- Soldering block or charcoal block

- Binding wire (optional)
- Paintbrush
- Torch for soldering
- Steel tweezers
- Reverse-action tweezers
- Tin snips
- Pickle in slow cooker
- Glass bowl with water and tongs (copper, brass or plastic)

FINISHING AND POLISHING TOOLS

- Abrasive paper
- Sanding stick (optional)
- Barrel polisher, brass brush or brass polishing wadding (Brasso)

MISCELLANEOUS ITEMS

- Kitchen paper
- Scouring pad (optional, for other textures)
- Masking tape
- Double-sided tape

MATERIALS

- 25 × 25mm (1 × 1in) of 0.8mm (20-gauge) thick silver sheet
- 0.8mm (20-gauge) wire to make one jump ring
- Wire for texturing – e.g. copper, silver or brass wire approximately 0.8mm (20-gauge) thick
- Easy solder

MAKING THE SILVER PENDANT

MARKING OUT A CIRCLE ON THE SILVER SHEET

Draw around a coin or choose one of the following two methods to draw a circle on the silver sheet.

USING DIVIDERS

Draw a circle approximately 25mm (1in) in diameter on the silver sheet. To do this, use dividers to draw the circle, as shown below.

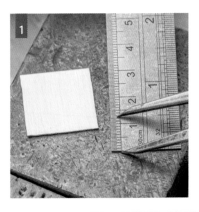

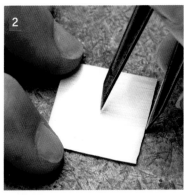

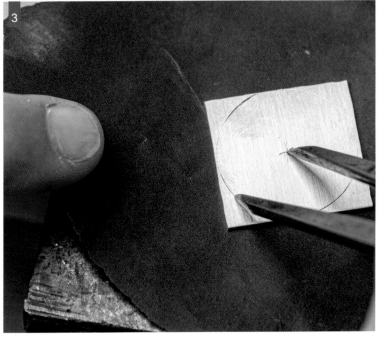

HANDY HINT

If your sheet is square, you can draw diagonals from corner to corner across the sheet with a pencil to find the middle. Use a pencil to avoid making marks across your sheet.

1. To find where the centre of the circle should be, set the dividers a little larger than the radius of the circle – 13mm (½in) in this case.

2. Place one leg of the dividers in the middle of one edge of the sheet and mark where the other leg touches the metal 90 degrees to that edge. Turn the metal 90 degrees and repeat so that the marks made on the metal cross. Reset the dividers to the exact radius (12.5mm in this case). Check that the mark is in the right place by placing one leg of the dividers on the centre mark and check that the other leg of the dividers doesn't fall off the edge of the sheet. If all is well, go on to mark the circle.

3. Place the piece of silver sheet on abrasive paper to prevent it slipping. Place one leg of the dividers on the mark you have made in the centre of the circle. Angle the dividers at approximately 45 degrees to the surface and turn the abrasive paper with the sheet on it away from the point of the dividers (don't work against the point or it will stick). Turn until you have a complete circle.

USING A CIRCLE TEMPLATE

As an alternative to dividers, use a 25mm (1in) circle template to mark out the circle to be cut out of the silver sheet (see left), using a scriber.

HANDY HINT

To see any scribed line more clearly, colour the sheet metal with permanent marker where the design will be before you scribe. Use any colour except black, as it will make it hard to see where you are sawing at the next stage. Let the ink dry before you start scribing.

CUTTING OUT THE CIRCLE

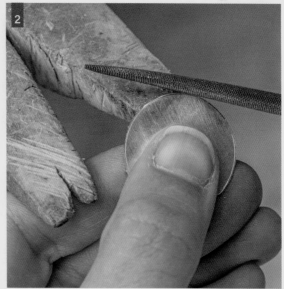

1. Using the piercing saw, cut out the circle. Do not cut on your scribe line – cut just outside it, so that when the circle is cut out you can still see the scribe line. This will enable you to file down to the line to make a neat circle. If you cut on the line you are likely to saw into the circle, which means you will end up having to draw another, smaller, circle inside the first one.

2. Using the flat side of a hand file, file down to your scribe line.

If there is a dot in the middle of your sheet from the dividers, either sand it out or simply texture the other side of the circle so that the mark is on the back.

TEXTURING THE SILVER CIRCLE

Choose two or three textures that you like, from the Texturing Metal chapter (see pages 104–111) and texture your circle. Below are the instructions for the textures we have chosen to apply:

HANDY HINT

If you are using a texture that affects the edge of the circle and you want a crisp edge, you can texture the metal first and then cut out the circle. Or, you can file the edge again after texturing the circle. If you do not want a crisp edge, you can leave the edges as they are.

The textured pendant after we have filed the hammered edge to make it crisp.

1. Use a hammer to texture the bottom of the circle (see page 104). Here, we have used a ball pein hammer.

2. Then hammer wire over the surface to add lines (see page 107) that create a seascape with waves and a seabird. We have used the flat face of the hammer to hammer the wires.

3. Texturing can make the silver sheet go concave. If this happens, turn it over, lay it face-down on a sheet of kitchen paper and use a mallet to flatten it.

HANDY HINT

Remove any burrs around the edge of your circle by sanding with a sanding stick or with abrasive paper wrapped around a file held at roughly 45 degrees to the edge.

DRILLING A HOLE FOR A JUMP RING

At this stage you will need to drill a hole for a jump ring to go through – the jump ring will form the **bail** of the pendant, through which a chain will go.

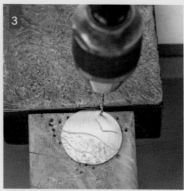

1. Make a gentle tap with a centre punch to mark where you want the hole to be. This will stop the drill bit from slipping across your metal. Lay the pendant on a steel block (such as your anvil) to ensure a good indentation and support the metal as you make your mark. The hole must not be too close to the edge of the pendant.

2. Move the pendant onto the benchpeg to drill. Before you begin drilling, you will need to prevent the metal from spinning around. Try any of the methods listed in the box below, Securing the Metal before Drilling.

3. Drill with a hand drill. Do not press too hard or the drill bit will break. When the drill bit is nearly through, it catches and you will feel more resistance and then it will go through the metal to the wood. Drill bits cut better at lower speeds.

4. There will be a burr on the back of the hole. If you have a larger drill bit, spin it in the hole by hand to cut off the burr. If you don't have a larger drill bit, sand off the burr.

SECURING THE METAL BEFORE DRILLING

Try any of these methods to secure your metal on the benchpeg:

a) Put double-sided tape on the back of the circle and stick it down on the benchpeg.

b) If you have a step between the benchpeg and the steel anvil, sit the circle on the benchpeg against the steel. This is shown in photograph 3, above.

c) Use a spare bit of wood and place it on your table against the anvil. Put your metal on top of this wood against the anvil and drill. The anvil will stop it spinning. The wood can be your drilling block to avoid putting holes in your benchpeg or table.

DOMING THE PENDANT

CHOOSING YOUR DOMING PUNCH

The correct doming punch is the one that goes into the recess with a little bit of space around it. It is important to choose the correct punch: one that is too small will not dome the piece properly; one that is too big will not reach the bottom of the recess. If the punch is the wrong size, it will get damaged.

HANDY HINT

Listen to the sound as you hammer the punch: you will hear a change in note when the metal hits the bottom of the recess. This is when you should stop. With the largest recesses, start doming the piece around the sides before working towards the middle. If you dome straight to the middle, the piece may crease.

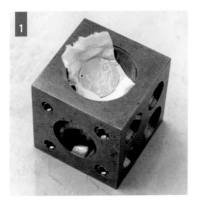

1. Decide whether you want your texture on the inside or outside of the dome. This decision determines if your pendant is convex or concave. For a convex dome, place the circle texture face-down in the block. For concave, place it face-up. To protect the texture, place either leather or kitchen paper between the textured side and the steel.

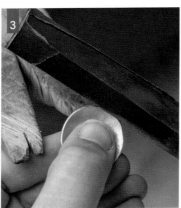

2. Dome the pendant using your largest doming punch in the largest recess. If you do not have the right size punch for the larger doming recesses, you can use a smaller punch with folded-up kitchen paper. Fold the kitchen paper four times so that it becomes a thick pad. Place the folded paper on top of your metal in the doming recess and hit the biggest punch that you have with a mallet to shape it into a dome. You will need to move the doming punch around in the recess to push the metal down and shape it. Keep checking the metal – if there are parts of the metal that are not becoming domed, repeat the process using the punch on those specific parts, still using the kitchen paper pad.

3. Finally, sand the edges of the domed pendant if needed.

MAKING AND SOLDERING THE JUMP RING BAIL

Make a jump ring with the 0.8mm (20-gauge) round wire that is big enough to go through the hole in your pendant and leave enough space for a chain to go through. (See How to Make One Jump Ring, page 88.) Remember that the chain ends will be wider than the chain itself. If any of the jump rings at the ends of the chain are unsoldered, you can take the ends off, thread the chain through the bail and then put the ends back on again.

HANDY HINT

If you make a heavy pendant to hang from the jump ring, it is best to support the weight of the pendant when soldering the jump ring. If it is not supported, the weight of a heavy pendant can pull open the jump ring while you are soldering.

1. Put the jump ring through the hole in your pendant and close the join for soldering using a pair of chain-nose pliers and a pair of flat-nose pliers.

2. Hold the jump ring with reverse-action tweezers so that the pendant hangs over the edge of the soldering block. Rest the reverse-action tweezers on the soldering block. Position the solder on the join of the jump ring. Easy solder is a good option

here because of its lower melting temperature, which reduces the risk of melting the jump ring.

3. Solder the jump ring closed. It is important to solder the jump ring so it does not open when you wear the pendant.

4. Pickle, rinse and dry.

FINISHING THE PENDANT

Polish the pendant with any of the previously mentioned methods. You can barrel-polish any textured options – however, if you have oxidized your piece do not use the brass brush, as it will remove the effect.

PROGRESSION

Squares and circles of sheet have been textured and then domed. They have been made into earrings using the methods from project 3.

Use the same method as in project 3 to melt a ball onto the end of a length of wire. Texture the wire with a hammer, anneal and use pliers to twist it into a swirl. Add a jump ring at the top to turn your piece into a pendant.

Textured rings made from either sheet or rectangular wire which have been textured while flat, before being made into rings.

Make pillow beads. Texture and then dome two of the same shape, such as squares, rectangles or hearts. Sand them on a flat surface to create flat points, so when you put them together this is where the solder join is. There is no need to make a bail as the chain can be threaded through the bead itself.

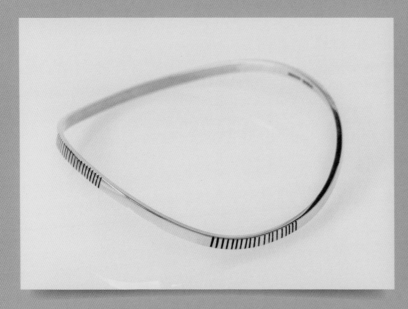

Use a piercing saw – see page 110 – to make lines on a bangle. Oxidize the lines to make them stand out – see page 111.

MAKING A BEZEL

A bezel is one way to secure a gemstone in a setting, as in the following project on pages 126–145. A bezel is the metal that goes around the stone to hold it in.

ABOUT BEZEL STRIP

Bezel strip is a convenient option to use to make a bezel. It is made from **fine silver** (99.9% silver instead of 92.5% silver, which is sterling silver) and is very soft and thin. Depending on your supplier, you can buy bezel strip in several widths and a couple of thicknesses. When setting larger (over 3cm – 1¾₆in – diameter) stones, use the thicker option. For the cabochon we recommend for the stone setting project that follows (see pages 126–145), bezel strip will work well. However, be aware that bezel strip will not suit all stones – see below.

IMPORTANT NOTE

Bezels need not be made out of bezel strip – you can cut your own from sterling silver and fine silver sheet. When using thicker materials than bezel strip, the resistance may be too much to push it over the stone by hand with a burnisher. You may need to use a punch with a hammer and other methods of supporting the work while setting the stone (such as a vice with soft jaws).

PROS AND CONS OF BEZEL STRIP

PROS:

- It is soft and therefore easy to rub over the stone;

- It is already the right thickness to rub over the stone;

- It already has straight edges;

- It is fine silver, so there is no firestain (see page 159);

- Because it is so soft, it is easier to shape around unusual-shaped stones.

CONS:

- If using 0.3mm- (28-gauge-) thick wire, there is not much thickness to file away marks and dents, so you have to be careful when working it;

- Using 0.3mm (28-gauge) thick wire is not the best choice for larger stones that are over 3cm (1¾₆in) in diameter) because its thinness makes it too flexible, making it harder to handle;

- The set widths available may not suit your stone. Adjustments may need to be made or it may not be high enough for your stone.

WORKING OUT THE LENGTH OF A BEZEL

When ordering the material for the bezel, it is useful to order more than you need. This allows for the supplier not cutting the ends square, the stone not being quite the size you ordered, errors in calculation and so on. Also, if you melt the whole thing, you have more material to make it again. Use one of these methods to determine the length needed:

1. Wrap the bezel strip directly around the stone. Make sure that the end is filed square first. The stone can be held down with double-sided tape if you find this easier. With the stone held down, form the bezel strip around the base of the stone until it overlaps. Mark where it overlaps and cut just outside your mark. Take care to keep the bezel strip vertical and not curve it over the stone. It needs to follow the shape of the base of the stone. This method has been used for the stone featured in the project (see pages 128–129).

2. Wrap a strip of paper around the base of the stone and mark it in the same way as above to give you the length of bezel you need. Instead of a strip of paper, you can also use a cut-down sticky note to wrap around the base of the stone. This technique is best used on stones above 8mm (5⁄16in) in diameter.

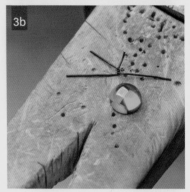

3. Wrap a thin wire, such as binding wire, around the base of the stone. Twist it together so it joins and fits snugly around the base of the stone.

a) Take it off the stone and cut it with wire cutters on the opposite side to the twist;

b) Open it up and straighten it and you will have the length for your bezel.

4. These formulae work out the length of bezel strip you will need. The formula adds the thickness of the metal, otherwise the bezel would be too small.

Round stone:

(diameter + thickness of the bezel) $\times \pi$

Oval stone:

(((long diameter + short diameter) $\div$ 2) + thickness of bezel) $\times \pi$

For example, in the project (see pages 126–145) we are using a 10mm-diameter stone with 0.3mm-thick bezel strip. Therefore, the calculation for the length of the bezel will be:

(10 + 0.3) $\times \pi$

That is, 10.3 $\times$ 3.14 = 32.342mm, which you can round up to 35mm (1⅜in) to allow for wiggle-room.

IMPORTANT NOTE

π or pi is the ratio of the circumference of a circle to the diameter of that circle. For our purposes, we can assume that π is equal to 3.14. It allows us to use the diameter of the stone to work out how long the bezel needs to be to wrap around it. You can use a scientific calculator or use the calculator app on a smartphone (many calculator apps show pi if you turn the screen on its side and view it in landscape mode).

GETTING THE CORRECT HEIGHT OF THE BEZEL FOR THE STONE

The bezel is at the correct height when there is a small gap between the top of the bezel and the curve of the stone. The bezel needs to be approximately 0.5mm above the start of the curve of the stone.

To see this, place your stone on a flat surface and place the bezel vertically against it. Look at where the top edge of the bezel comes on the stone. There should be a gap between the top edge of the bezel and the stone. It is this gap that is closed when setting the stone and stops the stone coming out. If you cannot see a gap, the bezel strip is not high enough. If the gap is very big, too much material will cover the stone and could crease when pushing it over the stone. This means the bezel is too high.

Do not assume that the size of the stone dictates the height of the bezel. A larger stone may have a shallow profile and a small stone may be very tall.

✗

The bezel is too low with no gap between the stone and the top of the bezel; therefore, there is no gap to close to hold the stone in.

✓

The bezel is the correct height with a gap between the top of the bezel and the curve of the stone. This will be closed to hold the stone in.

✗

The bezel is too high. When set, too much of the stone would be covered and you risk the metal wrinkling.

ADJUSTING THE HEIGHT OF A BEZEL

If a bezel is too low for a stone, another needs to be made as it will not hold the stone in securely. If a bezel is too high for the stone, you can either raise the stone or lower the bezel.

RAISING THE STONE

To raise the stone to the correct height, you can choose to make a booster seat (see page 136). A booster seat is a ring of wire that lifts the stone to the correct height in relation to the bezel. Choose a thickness of wire that will lift the stone up to the right height: you may not get the height right the first time, but as the wire is not soldered in, you can remove it and try again (with thinner or thicker wire as needed).

The stone can also sit on fine sawdust to lift it up in the stone setting (see page 143).

LOWERING THE BEZEL HEIGHT

Choose any of these three options to lower the height of the bezel:

1. If the setting is too high, it can be sanded down. Do the sanding after you have soldered the bezel closed, and use a sanding stick or abrasive paper that is supported on a flat surface. Sand in a circular or figure-of-eight motion so that it is not lopsided. Check that you have sanded enough by pushing the bezel back over the stone. Remember that getting the stone out may be difficult if it is a tight fit.

2. The height can be filed down. Do this after the bezel has been soldered closed. It helps to use dividers to mark where you need to file down to so that you can check that you are filing straight. Be careful when you file so that you don't bend the bezel. Use a fine file.

3. If the bezel strip is too high to sand or file to the correct height, it can be cut before you form it.

a) Scribe a line with dividers along the bezel strip at the height that you want.

b) Use tin snips or good-quality scissors to cut just outside the line. It might be easier to draw two lines and cut in between to get the line as straight and level as possible.

Be aware that the cut edge stretches and can curl the bezel strip. This can be corrected after you have soldered it shut by tapping it level with a mallet on a steel block.

PROJECT 5:
STONE SETTING FOR A RING

The finished stone setting mounted on a ring.

WHAT WE ARE MAKING

A stone setting for a cabochon stone to put on a ring that you have previously made.

Stone setting is the process of securing a gemstone in jewellery after all heat processes are finished. Most stones cannot survive the soldering process, so the fabrication of the piece must be completed before the stone is set. This type of setting is called a rub-over setting, as the metal is rubbed over the stone to hold it in.

CORE SKILLS

Sawing; filing; forming; sanding; soldering; pickling; polishing.

NEW SKILLS

Making a rub-over stone setting; making a bezel; setting a cabochon stone; soldering the stone setting onto a ring; making a booster seat.

YOU WILL NEED

BENCH, HAND AND FORMING TOOLS

- Needle files
- Hand file
- Scriber
- Piercing saw and saw blades
- Beeswax or candle wax
- Wire cutters (optional)
- Tin snips
- Ring clamp (optional)
- Half-round pliers
- Chain-nose pliers
- Hide mallet
- Burnisher

SOLDERING TOOLS

- Flux (such as borax) and dish or piece of slate
- Large soldering sheet
- Soldering block or charcoal block
- Paintbrush
- Torch for soldering
- Steel tweezers
- Reverse-action tweezers
- Tin snips
- Pickle in slow cooker

- Glass bowl with water and tongs (copper, brass or plastic)
- Steel mesh (optional)

FINISHING AND POLISHING TOOLS

- Abrasive paper
- Sanding stick (optional)
- Felt stick with Tripoli (optional)
- Barrel polisher/brass brush or brass polish wadding (Brasso)

MISCELLANEOUS ITEMS

- One of the rings you have previously made – see pages 60–87
- Double-sided tape
- Dental floss

MATERIALS

- Hard, medium and easy solder
- Stone for setting: we are using a 10mm (⅜in) round cubic zirconia **cabochon** stone
- 35mm (1⅜in) of 3mm (⅛in) wide × 0.3mm (28-gauge) thick bezel strip
- 12 × 12mm (½ × ½in) piece of 0.5mm (24-gauge) thick sterling silver sheet
- A length of your 0.8mm (20-gauge) sterling silver round wire for the booster seat

MAKING THE BEZEL

SIZING AND FITTING THE BEZEL STRIP

In this project, you will be making a bezel that is the correct height for a 10mm (⅜in) cubic zirconia cabochon stone. The first step is to make the bezel strip fit around your cabochon. The bezel strip is thin and flexible, so it needs to be handled carefully. Repeated bending will result in wrinkles. Removing these wrinkles by filing or sanding could result in the file going through the metal, so avoid making the wrinkles in the first place.

HANDY HINT

See page 123 for the calculation of the length of your bezel for this project.

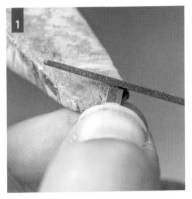

HANDY HINT

If you find that the stone moves around too much when you are shaping the bezel strip around it, use double-sided tape to hold it down on your work surface.

1. File the end of the bezel strip so it is straight and at a right angle to the long edge. To file, hold the bezel wire very close to the end so it does not bend, and file along the length, rather than across it, for the same reason.

2. Place your stone on a hard, flat surface. Wrap the bezel strip snugly around the base of the stone. Make sure that you don't push it over the curve of the stone – keep it vertical.

3. Mark with a scriber exactly where the filed end overlaps the other end of the bezel strip.

4. Remove the bezel strip from the stone. Cut outside and right next to your scribe line so you can file down to it. Use one of the methods described in the box opposite to cut.

CUTTING BEZEL STRIP

To cut with a piercing saw, straighten the bezel strip where the mark is so you can lay it flat on the benchpeg. Saw straight down onto the benchpeg across the width of the bezel strip. You need to cut this way because the wood supports the bezel strip and the blade will run smoothly along its width.

To cut with wire cutters, make sure that you cut it square to the edge. If using this method, the wire cutters need to be good quality to cut through cleanly. If you are unsure they will cut cleanly, test your wire cutters on a scrap piece first.

If cutting with tin snips (as shown in photograph 4, opposite) or good-quality scissors, make sure that you stay square to the edge. It will put a curve across the width of the bezel, which will be corrected in step 7.

5. Once cut, check that the size of the bezel is correct by wrapping it around the stone again. The ends need to butt together and hold the stone. If it is too loose, trim the bezel again. If it is slightly too tight, it can be stretched later. If the bezel strip is more than 2mm (1⁄16in) too short, start again.

6. When you have the right length, file the cut end as in step 1. This is important as the bezel strip is only 0.3mm (28-gauge) thick and filing makes sure that there are two 0.3mm (28-gauge) edges coming together, not the pinched edge you get if you used wire cutters or tin snips to cut the bezel strip.

7. To prepare the bezel strip for soldering, bring the ends together so they line up and there is no gap. Using flat- or chain-nose pliers, squeeze gently over the join from one side and then from the other side. This lines up the join exactly and also gets rid of the curve in the metal if you have used tin snips to cut it.

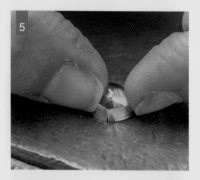

SOLDERING THE BEZEL

1. Place a pallion of hard solder on the soldering block.

2. Place the join of the bezel on top of the pallion of solder with the join facing you.

3. Heat the block in front of the join and move the flame towards the join. When the solder flows, the bezel will drop down onto the block. Move the flame away.

4. If the solder has flowed only on one side of the join, repeat the above with a new piece of solder.

5. Pickle and rinse. Do not file or sand the join at this stage. It is good to leave the extra solder on so there is less risk of the join emptying out when soldering the bezel to the back sheet.

SHAPING AND SIZING THE BEZEL

Above, a perfect fit on the left; an incorrect fit on the right – the bezel is too big for the stone and does not shape to it; there are gaps between the bezel and the stone.

1. Push the bezel straight down over the stone to shape it. As the bezel strip is soft, it will conform to the shape of the base of the stone.

2. Take the bezel off, turn it over and push it over the stone again. If the size of the bezel is correct, it should fit over the stone. When you slide the stone and bezel back and forth over a table, the stone should not move side to side inside the bezel. Check the shape and size of the bezel by turning it over with the stone in. If you can see gaps between the base of the stone and the bezel, it is too big. If it is too small, the bezel will not push down to the base of the stone. (See Troubleshooting, below.)

TROUBLESHOOTING

THE SIZE OF THE BEZEL

PROBLEM

The bezel is too big.

SOLUTION

Cut open the solder join. Wrap the bezel around the stone again, taking care to keep it vertical. Mark the overlap and cut off the excess. Remember to file the ends flat again before soldering.

PROBLEM

The bezel is too small.

SOLUTION

Stretch it by squeezing with pliers. Use half-round pliers with the curve on the inside. Squeeze three times, moving along the bezel. Repeat on the other edge to avoid making a taper on the bezel (i.e. making it cone-shaped). Try the fit again over the stone and repeat until it fits. Remember, if it is much too small you will need to start again. The pliers will make a slight ripple on the edge of the bezel strip, so sand the top and bottom edges flat.

SOLDERING THE BEZEL ONTO A BACK SHEET

The bezel is soldered onto a back sheet in order to stop the stone falling through the setting. Use the 12 × 12mm (½ × ½in) piece of 0.5mm (24-gauge) thick sterling silver sheet for the back sheet.

HANDY HINT

When soldering, start by placing the join of the bezel facing away from you. This protects the solder join by starting the solder flowing on the opposite side, holding the bezel in place: if you then overheat the bezel join, it is less likely to spring open.

SAFETY NOTE

When you use tweezers to lift up the metal, remember that they will get hot. Either leave them to cool or pick them up with other tweezers and quench in water.

1. Sand the top and bottom edges of the bezel on abrasive paper. The paper needs to be on a flat, solid base, such as your work surface or a sanding stick. This makes the edges flat and prepares the bezel for soldering.

2. Sand the surface of the sheet that the bezel will be soldered on, to clean it for soldering.

3. Place the bezel on the back sheet and check that there are no gaps. Sand the bezel strip more if there are gaps, in order to remove them.

4. Flux the top of the sheet and place the bezel on it while still wet. This will also flux the bottom of the bezel.

5. Two options to set up the soldering are:

a) If you have a steel mesh, place the piece on it.

b) Place the piece on the soldering block with tweezers underneath it to lift it off the soldering block. The tweezers have been laid on their side and are not being held.

6. Put pallions of medium solder on the sheet against the outside of the bezel. In photographs **5a)** and **5b)**, above, you can see the pallions in place.

7. As the sheet is bigger than the bezel, it needs more heat to reach soldering temperature. Both setting-up methods shown above allow the heat to go under the back sheet so that it heats up first. Enough heat will conduct up to the bezel, heating this too, but protecting the bezel from overheating as you do not have to point the flame directly at the bezel.

8. The solder should run right around the whole bottom perimeter of the bezel. If the solder has run on only one side, it is usually the side towards the flame where it is hotter. Turn the whole thing around so you can heat the side where the solder has not flowed.

9. Check that the solder has run all the way around the base of the bezel. If it has not, you must fix it at this stage. You may need to add more solder.

10. Pickle and rinse.

TROUBLESHOOTING

SOLDERING THE BEZEL

PROBLEM

There is a gap in the join of the bezel before it has been soldered to the back sheet.

SOLUTION

Place the saw blade against the join and cut it open. When cutting the join open, as you can see in the photograph below, the blade will cut into the wood before reaching your fingers. This will remove the gap and the join will be ready to be soldered again. This could make it too small so, to remedy this, see page 130.

PROBLEM

There is a gap in the join of the bezel after it has been soldered to the back sheet.

SOLUTIONS

1. Try to fill the gap with medium solder. As soon as the solder starts to flow, remove the flame or it will flow away from the gap.

2. If this doesn't work, place a length of thin wire along the join. The flux will hold it in place. Do not use too much or too-thick flux, as it may dislodge the wire as you heat. Use medium solder to solder it. The solder will run under the wire, thus filling the join of the bezel. This wire is sacrificial and will be filed away. 0.8mm (20-gauge) round wire is good for this method.

If neither of these methods works, start again.

PROBLEM

There is a gap between the bezel and the back sheet.

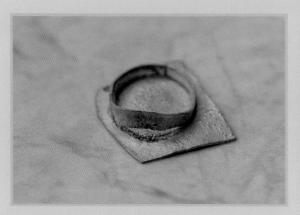

The gap in this sample is too big to fix. It is not worth repairing as it has been so overheated that the back sheet has started to melt, which you can see by the texture.

SOLUTIONS

- If the bezel is not fully soldered, try flowing more medium solder around the base of the bezel where the solder had not flowed before.

- If the back sheet has slumped, push up the back sheet with your fingers to close the gap. Then try to flow medium solder around the base again.

- If the bezel has lifted away from the back sheet, push it down with your fingers to close the gap. Then try to flow medium solder around the base again.

- If this does not work, make the setting again. If the bezel is soldered to a larger piece, you can remove it. To remove the bezel from a piece, reflow the solder and, when it is flowing, pull the bezel off with tweezers. This will not salvage the bezel, but at least you will have the back sheet.

PROBLEM

The solder has not run all the way around the join.

SOLUTION

Flux again and add another piece of medium solder where it has not flowed the first time.

TRIMMING AND FILING THE SHEET DOWN TO THE BEZEL

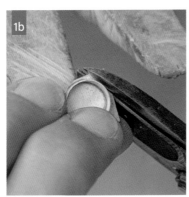

1. Cut off the excess sheet around the bezel using one of these methods:

a) Using a piercing saw, cut around the bezel taking care to keep the blade upright so you don't cut into the bezel. Cut as close as you can, but not against the bezel.

b) Using tin snips, trim the sheet from around the bezel. Again, cut as close as you can, but not against the bezel. It can be easier to do this in straight lines rather than trying to go around in a circle. Do be careful not to get your fingers caught in the jaws or the stop on the handles.

2. File the sheet down to the bezel. You should see the shiny file marks on the edge of the sheet only, not on the bezel. To support the piece while filing, hold the back sheet against the benchpeg – do not push the bezel against it. Keep filing until the file marks just start to touch the bezel. You know you have filed far enough when you cannot see any ledge.

3. Tidy up the join on the bezel strip using a hand file or needle file. If you find that there is an indentation there and you cannot file without making it too thin, wait to do this step after you have set the stone. Setting the stone may help to push out the indentation.

4. Sand out the file marks using a sanding stick or abrasive paper wrapped around a file.

HANDY HINT

There is no need to put the stone in the bezel to test the fit; however, if you must, we strongly advise that you lay a length of dental floss over the bezel then put the stone on top of it so that the stone is easy to lift out. Refer to Troubleshooting, page 138, if the stone gets stuck.

SOLDERING THE STONE SETTING ONTO THE RING

Setting up the ring with a pallion of solder against the shank.

1. Choose one of the rings you made in a previous project. Sand the top surface of the ring opposite the solder join to prepare for soldering. If you are using a twisted wire ring, you will need to file a flat area for soldering.

2. Sand the underneath of the stone setting. Use fine abrasive paper so that you are not adding scratches that need to be removed later.

3. Place the bezel upside-down on the soldering block and flux where the solder will go.

4. Flux the ring where it will touch the bezel.

5. Hold the ring on one side with the very end of the reverse-action tweezers going through the ring.

6. Place the reverse-action tweezers with the ring on the soldering block so the ring rests on the back sheet. Make sure that the ring is vertical to the back sheet of the bezel. Remember to place the solder join of the ring opposite where you are soldering the ring onto the back sheet.

7. Place a large pallion of easy solder on the back sheet against the **ring shank**. The shank is the part of the ring the finger goes through.

8. Dry the flux with the flame, then heat the ring first and move the flame down to the solder. Avoid heating the bezel strip by keeping the flame directed on the back sheet and the ring. Heat until the solder flows.

9. Pickle and rinse.

Setting up the ring with solder already melted on the shank.

HANDY HINT

You might like to try this alternative approach to soldering a ring shank to a setting, to see which you prefer. Opposite the solder join on the ring, flow a piece of easy solder. Then set it up as in the project, with the easy solder now between the ring shank and the setting. Reflow the easy solder to join the two together.

MAKING A BOOSTER SEAT

The 3mm (⅛in) bezel strip will be too high for the stone we have used in this project, so it will need to be lifted, or boosted, in its setting. To do this, you can make a booster seat out of wire, on which the stone will sit.

The booster seat needs to be tight enough so that it does not rattle from side to side in the bezel, but it does not need to be an exact fit. It will not be soldered closed or soldered into the setting. No part of the booster seat should be higher than another – the seat should be level and sit flat.

HANDY HINT

A booster seat can be used for aesthetic, as well as practical, reasons. If you would like your stone to sit high in your design, or you want the stone to look a bit deeper than it actually is, it can be lifted in a taller stone setting using a taller booster seat.

Using half-round pliers, make a round ring out of the 0.8mm (20-gauge) round wire that will fit inside the setting. The booster seat should sit flat to make a level seat.

The booster seat inside the bezel setting.

SETTING THE STONE

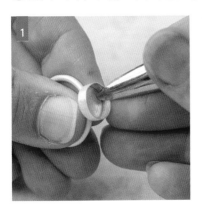

1. As the bezel is soft, it may have bent inwards, making it too small to put the stone in. If this is the case, run a burnisher or half-round pliers around the inside of the bezel to push the wall back out to a vertical position again. Make sure that you push out the bezel all the way from the bottom, not just the top edge.

HANDY HINT

When setting the stone, you can hold the ring in a ring clamp. Put the ring shank all the way in so the bottom of the setting is resting on the ring clamp jaws.

2. Put the stone upside-down on the work surface and put the booster seat on top of it. Putting it upside-down means it is easier to lower the bezel straight so that the stone doesn't go in crooked. Lower the bezel over them both to check the fit of the stone. Do not push it if it is tight. Instead, lift up the ring and repeat step 1 to loosen it a bit more. Once the bezel slides snugly over the stone and booster seat, push the ring down onto the stone so that the stone is level and all the way in.

3. Turn the ring over; hold it securely between your index finger and thumb and hold the burnisher in your dominant hand. To stop the burnisher from slipping while setting, hold one thumb on top of the other to anchor your hand.

4. When setting the bezel, work on opposite sides at a time (think north-south-east-west) to keep the stone in the centre of the setting. Start by pushing the bezel from the bottom to the top of the setting using a rocking motion. Don't push the bezel completely over the stone the first time around. Move to the opposite side and do the same again. Now go in between these two points and do the same again. Keep doing this, continually working on opposite sides until you have worked all the way around the bezel.

5. The second time around, still working on opposite sides of the bezel, push it completely down against the stone. Make sure to start at the bottom of the bezel because the bezel needs to be in contact with the stone from bottom to top, not just the top edge.

6. To smooth the bezel against the stone, rub the burnisher around the bezel against the stone and towards your body. This will smooth out the bezel and help to push it down against the stone.

7. Finally, to highlight and smooth over the top edge of the bezel, use the side of the burnisher to rub around the top thickness of the bezel. Remember to anchor your hands together to have better control and also work a small section at a time. Both of these methods will prevent the burnisher slipping.

ABOUT THE BURNISHER

A burnisher is one of many tools that can be used for setting stones. You can also use a rocker, a pusher or even the back of the bowl of a steel teaspoon if you are using bezel strip. The benefit of a burnisher is that it can also be used for other things, whereas a pusher or a rocker are used only for setting.

TROUBLESHOOTING

SETTING THE STONE

PROBLEM

The stone needs to be removed from the setting for some reason. This can be because the stone has gone in crooked or you notice another problem that needs heat to be resolved.

SOLUTIONS

If you have not already started to set the stone:

a) You may be able to remove it by sticking duct tape on the stone to pull it out.

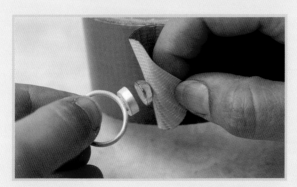

b) Alternatively, you may be able to remove it by putting the piece in a cardboard or plastic box with a lid and shaking it until the stone is released.

If you have started setting the stone:

c) You may be able to remove it by gently prising back the bezel with your thumbnail.

d) You may be able to remove it by pushing back the setting with a tool. In the photograph below, we have used a dental probe that has a flattened end. You can file the end of a bit of sheet so it is thinner and use that to prise it away from the stone. Avoid damaging or bending the bezel as much as possible if it is going to be used again. Continue by using options a) or b).

PROBLEM

The join of the stone setting – where it is attached to the jewellery – breaks while setting the stone, so the setting comes off the jewellery with the stone in it.

SOLUTIONS

- If you can remove the stone and salvage the setting, do so and resolder the setting back onto the work (see page 135).

- If you cannot get the stone out without damaging the setting, you will have to cut the stone out of the setting and remake the whole setting. Use a piercing saw to cut through the side of the bezel strip in two places so that you can peel it back. Make a new setting. If the stone is above 7.5 on the Mohs scale of hardness, the saw blade will not damage it – in fact, the stone will blunt the blade.

PROBLEM

When you have set the stone, it still moves or wiggles, so the stone is not fully set.

SOLUTIONS

- Use a burnisher to tighten the setting again – the bezel may not have been fully pushed against the side of the stone. Remember that you have to push the bezel over the stone from the bottom to the top. This is especially important for round stones, as they can spin in the setting if you push it over only at the top.

- You may not be strong enough to push down the bezel securely with only a burnisher. You can use a punch with a flat end and tap – push – down the bezel strip by gently tapping the punch with a hammer or mallet.

- The bezel may have been too big when it was made. When setting, if you see that the bezel strip is bending in at the bottom over the back sheet before it gets to the stone, your setting is too big for the stone and you will need to start again.

FINISHING AND POLISHING

REMOVING MARKS

If there are any marks made while setting the stone or if the solder joins show after soldering the bezel to the ring, now is the time to clean them up, before you polish your ring.

If you have put in deep marks, you may need to remove these with a file and then sand out the file marks. Files (other than diamond files) are 6–7 on the Mohs scale of hardness (see pages 190–191), so will not mark harder stones. Abrasive paper is 9 on the Mohs scale of hardness, so will mark all stones except diamond.

When using abrasive paper, put your thumbnail over the stone to protect it while removing marks and polishing, as the paper will scratch the stone.

You can use a felt stick (buff stick) with Tripoli compound to avoid using abrasive paper to polish your bezel.

1. If you can see the solder join on the bezel, concentrate on this first. When sanding with the stone in the setting, protect the stone – put your thumbnail over the stone, put masking tape over the stone or simply avoid touching the stone with the abrasive paper. File the join first, if necessary, or go straight to using a sanding stick.

2. If there are marks to sand off on the bezel strip, do this, being aware that the bezel strip is very thin, so do not sand too much or you will go through it.

3. You can avoid using abrasive paper near the stone by using a buff stick (polishing stick) with Tripoli polishing compound instead (see page 22). If your stone is 7 or above on the Mohs scale of hardness, Tripoli will not mark it.

Your stone-set ring is now ready to be polished.

POLISHING

You can barrel-polish before or after setting depending on the stone used. As a general rule, if the stone is hard or not brittle (7 or above on the Mohs scale of hardness) it can be set first and then barrelled. If the stone is softer, the piece is barrelled first and then the stone is set. However, if the stone has flaws in it, do not risk barrel-polishing it. If your stone is softer, such as turquoise, pearl, opal, fluorite or lapis lazuli, the surface will be ruined by the barrel polisher. Brittle stones such as emerald and apatite will also be damaged in the barrel polisher.

Be aware, if you barrel-polish your setting before setting the stone, the setting will be hardened slightly by the barrelling.

If you are unsure that the stone will survive barrelling, you can barrel-polish a spare stone on its own to test it. If you don't have a spare stone, barrel-polish the piece first and then set the stone.

The advantage of setting first and then barrelling is that if you make any marks on your piece when setting the stone you can remove them before barrelling.

If you don't have a barrel polisher, use a brass brush (see photograph above, right) or polish with metal polish wadding, such as Brasso. If you have used the square-wire ring from project 1, highlight the edges of your ring with a burnisher (see Texturing Metal, page 110).

PROGRESSION

Solder a bezel strip onto a bigger back sheet to create a decorative edge. Add it to a ring shank to make a statement ring.

Combine a stone setting with earring hooks made following the instructions in project 3 (see page 99) or using a wire wrap loop as in the example above.

Combine a stone setting with twisted wire from project 2 to make earrings.

Put your stone setting in a domed circle pendant based on project 4.

Beyond this project, there are many shapes and sizes of cabochon stone to use in your jewellery. The following pages will give you a basic understanding of the stones and designs you can use in a rub-over setting.

If you find that setting stones in jewellery becomes your interest, there is a whole world out there to investigate.

FURTHER CHOICES FOR STONES

SIZE:

If you are new to jewellery making and stone setting, it is best to start with cabochon stones that are 8–12mm (⁵⁄₁₆–½in) in diameter. When starting out, avoid using stones that are smaller than 6mm (¼in) in diameter, as they are fiddly to work with.

SHAPE:

Start with round or oval stones, and move onto other shapes as you gain experience. If you find a stone that you want to use but which does not sit flat, try these options when setting the stone:

- Put a little sawdust in the bottom of the setting to create a seat that stops the stone from rocking. The added advantage of sawdust is that, when you set the stone, the sawdust is compressed as you push the setting over the stone and when you release the pressure the sawdust expands, pushing the stone up tight against the setting.

- If the stone is domed at the bottom, you can put a ring in the setting like the booster seat (see page 136) for the stone to sit on to stop it rocking.

- If the stone is domed or bulbous at the bottom, you can cut out a correspondingly shaped hole in the back sheet. The stone will settle into the hole, holding it steady. This does mean the stone will be poking out of the back of the setting: consider the practicality of the piece you are making if you use this option.

This photograph shows a variety of cabochon stones of different shapes.

PROFILE:

The profile is the shape of the stone from the side. Some stones have a very high profile and some have a very shallow profile. It is best to gain experience with stones that have a regular curved profile. Don't start with stones that are very shallow or ones that have corners that lead to a flatter top (examples of these are the dark blue stones in the photograph below).

HARDNESS:

Stones are ranked on the Mohs scale of hardness (see pages 190–191). Start with stones that are a 7 or above on the scale, as there is less risk of damaging the stone when making the setting, setting the stone and polishing.

SHAPING BEZELS AROUND IRREGULAR STONES

If you are making a setting for an oval stone, wrap the bezel strip around the stone the same way as for a round stone.

If the stone has rounded corners, proceed in the same way as you would for a round or oval stone, but ensure that each corner is fitted correctly before you move on to the next, so that you do not get baggy areas where the bezel is too big. Ensure that the solder join is not at a corner, otherwise it will be more difficult to set the stone as the solder will harden that area. In addition, if you put the solder join at the corner, it will make a sharp corner as opposed to a rounded one, which will not look the same as the other corners.

If the stone has sharp corners, pull the bezel strip around to the first corner. Use flat-nose pliers to crisp up the corner. Then continue around and do the same at each corner. Don't put the join on or near the corner, as the solder will again make the corner harder and more difficult to set. Use this same approach for teardrop or pear-shaped cabochons, using flat-nose pliers to form the corner of the bezel that goes around the point of the teardrop.

If the stone has sharp corners, pull the bezel strip around to the first corner. Use flat-nose pliers to crisp up the corner.

SETTING AN IRREGULAR STONE

For bezels with corners, start by pushing the bezel at the corners. The first time, do not push the metal all the way down to the stone, but work gradually. Do each corner first and then work towards the middle of the sides. Doing the corners first avoids having too much material at the corners to push over the stone. Keep repeating until the stone is set. Be careful not to put too much pressure on the corners, as the stone can break more easily there.

For any bezel with corners, it can help to file down the height of the corners of the bezel a little so that when you set the stone there is less metal at the corners to push over the stone.

This photograph shows the corners of the bezel filed down.

CHOOSING THE BACK SHEET

If the stone is small, use thinner sheet – approximately 0.5mm (24-gauge) – for the back sheet so there is less filing down required after soldering.

If the stone is larger – above 2cm (13⁄16in) in diameter – it is best not to use sheet thinner than 0.6mm (22-gauge), as thinner sheet can warp when heating.

There are various design options available when designing the back sheet:

- The sheet can be cut down to the edge of the bezel after it is soldered onto it, as in this project.

- The sheet can be left bigger than the bezel and be cut to any shape you want. It can be best to cut the shape first and then solder on the bezel, but there are no hard and fast rules about which way around it is done. Textures can be made on the back sheet that show outside of the bezel, adding to the design. Whether you texture before or after soldering the bezel on depends on the type of texture chosen. (See the first progression piece on page 142.)

- A design can be cut into the sheet so the stone can be seen through it from the back. This can be as simple as leaving a ledge for the stone to sit on, or an elaborate design, such as flowers or stars. If the design is elaborate, it might be easier to saw it before soldering on the bezel, but there is no hard rule about the order. However, if cutting a ledge, this is best done after the bezel is soldered on, as it is easier to mark the shape being cut using the bezel as a guide.

- The bezel can also be soldered straight onto the jewellery piece, such as a pendant. Therefore, the piece becomes the back sheet.

A design can be cut into the sheet so the stone can be seen through it from the back. From left to right: a ledge has started to be sawn, a design has been cut out of the back sheet, and a flower has been cut out of the back sheet.

The bezel can be soldered onto a pendant, so the pendant becomes the back sheet.

The finished brooch.

WHAT WE ARE MAKING

A layered sheet-silver brooch using cut card technique.

Building on the skills you have gained from the previous projects, we will now make a brooch using cut card technique. Cut card technique is so called because the process is akin to cutting silhouettes out of card and gluing them together.

CORE SKILLS

Sawing; filing; texturing; sanding; soldering; pickling; polishing.

NEW SKILLS

Transferring a design onto silver sheet; sweat soldering; making a brooch **finding**.

YOU WILL NEED

BENCH, HAND AND FORMING TOOLS

- Piercing saw and saw blades
- Beeswax or candle wax
- Needle files
- Hand file
- Scriber
- Dividers (optional)
- Steel hammer
- Centre punch
- Hide mallet
- Round-nose pliers
- Flat-nose pliers
- Chain-nose pliers
- Wire cutters
- Ruler

SOLDERING TOOLS

- Flux (such as borax) and dish or piece of slate
- Large soldering sheet
- Soldering block or charcoal block
- Paintbrush
- Torch for soldering
- Steel tweezers
- Reverse-action tweezers
- Tin snips
- Pickle in slow cooker
- Glass bowl with water and tongs (copper, brass or plastic)
- Steel mesh (optional)

FINISHING AND POLISHING TOOLS

- Abrasive paper
- Sanding stick (optional)
- Felt (buff) stick with Tripoli (optional)
- Burnisher or steel teaspoon
- Barrel polisher/brass brush or brass polish wadding (Brasso)

MISCELLANEOUS ITEMS

- Paper for template
- Tracing paper
- Double-sided tape
- Masking tape
- Ruler
- Fineliner pen
- Scissors (optional)
- Kitchen paper

MATERIALS

- 6 × 4cm (2⅜ × 1½in) of 0.6mm (22-gauge) silver sheet
- 6cm (2⅜in) of 0.9mm (19-gauge) round silver wire
- 5mm (³⁄₁₆in) of silver tube with an outside diameter of 1.6mm and an inside diameter of 1mm. If you cannot get this exact size, go up one size rather than down in size.
- Copper, brass or binding wire for texturing
- Hard and easy solder

PREPARING THE BROOCH DESIGN

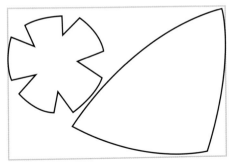

Above, the templates for the two components of the brooch design: the flower and the triangle. These are reproduced here at full size for your use.

HANDY HINT

Do not cut the designs out of the tracing paper to transfer onto the sheet, as this adds more inaccuracies. Leave the paper whole and tape the whole piece down and then cut through the tracing paper and the sheet at the same time, as shown on page 150.

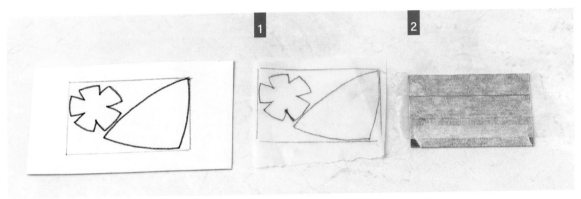

1. Using the templates provided at the top of this page, trace the triangular shape and the flower onto tracing paper.

2. Remove the protective plastic film from one side of the silver sheet. Put double-sided tape over the sheet on the side from which you have removed the protective film.

3. Stick down the tracing paper with the design onto the double-sided tape. If the tracing paper exceeds the edge of the sheet, you can trim the paper to size with scissors.

TRANSFERRING DESIGNS ONTO METAL

Tools and materials for transferring your designs onto metal.

Here are three methods from which to choose to transfer designs onto sheet metal:

1. STICKING THE DESIGN ONTO METAL

You can trace your design on tracing paper (as seen on the opposite page) and stick it down with either double-sided tape, glue stick, PVA glue or spray adhesive. Tracing paper will cut cleanly with the piercing saw and will help you to see where to place the design economically on the sheet.

2. USING CARBON PAPER

a) If you wish to use the same design again, you can keep the artwork by putting carbon paper between the design and the metal. Go over the design with a pen, leaving a carbon drawing on the metal surface.

b) Go over the carbon drawing with a scriber: when you hold the metal to cut out the design, your fingers can rub off the carbon lines. Check that the carbon paper is leaving a mark, because not all types of carbon paper will work.

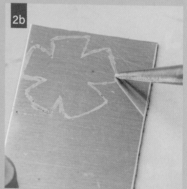

3. DRAWING DIRECTLY ONTO THE METAL

Drawing with a scriber, you can use templates, rulers, tri-squares or around anything that fits your design, such as a coin. Dividers can also be used to draw circles (see page 114).

Drawing onto the metal using a scriber and a circle template.

CUTTING OUT THE BROOCH DESIGNS

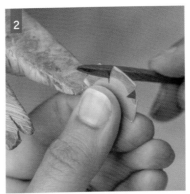

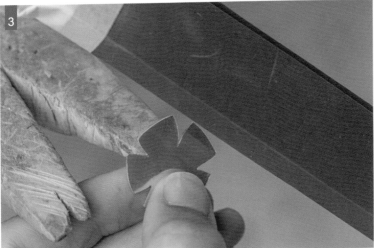

1. Using a piercing saw, cut out the two shapes, sawing through the tracing paper and the silver sheet at the same time. Remember to saw just outside the lines. Refer back to the tips on sawing corners in the Core Skills chapter (see page 37).

2. For both the triangle and the flower, clean up the edges and shape them with filing. Needle files are best to reach into the tighter areas. You can do this with the tracing paper on the sheet, or you can remove it and file the design by eye.

3. Remove the file marks on the edges of the flower and triangle using abrasive paper wrapped around a file or a sanding stick (see page 55).

If the filing has created a burr on any of the edges, remove it with a file or a sanding stick. Hold the file or sanding stick at roughly 45 degrees to the edge to remove the burr.

TEXTURING THE BROOCH

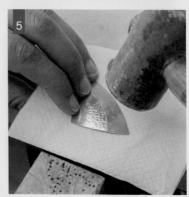

1. To make the lines on the triangle, curve two pieces of wire similar to the curves in the photograph to make the lines on the metal. If you have copper, brass or binding wire, use those. If not, use the 0.8mm (20-gauge) silver wire. Pull the wire through your fingers to remove kinks and put a curve on the wire.

2. Place the sheet on a metal block and tape the two wires down with masking tape – see images **2a)** and **2b)**.

3. Use the flat face of a general-purpose metal hammer to hammer the wires down to make a mark in the sheet (see page 107). Do not hit too hard or you will go through the sheet. Remove the wire. Use a centre punch to make dots in between the two lines. These do not have to go under the flower.

4. Use a centre punch to make the dots in the centre of the flower.

5. After texturing, the metal will curve up. Place the triangle face-down on a folded sheet of kitchen paper, on a steel block. Use a mallet to make it flat. Repeat with the flower.

The pieces are now ready to be sweat soldered together.

SWEAT SOLDERING

Sweat soldering is a way of soldering two pieces together by melting solder on one piece, placing the two pieces together and then reflowing the solder. The solder is usually melted on the back of the top piece so when the solder is reflowed it follows the shape of the top piece, meaning there is no excess solder to clean up.

Sweat soldering is the best method for soldering pieces of sheet one on top of the other.

HANDY HINT

Experience tells you how much solder to put on. If you think of the pallions as water running over the surface, have you got enough to cover the surface when it runs?

1. Before soldering the pieces together, remove any scratches from the top surface of the bottom piece (the triangle) first. This is because it is hard to clean around the top piece (the flower) once it is soldered on, which can make the surface of the triangle uneven.

2. Sand the bottom surface of the flower so it is clean for soldering.

3. Hold the two components together and mark on the reverse of the flower with a fineliner pen where it overhangs the triangle.

4. Put the triangle the right way up and the flower upside-down on a soldering surface. Flux both pieces.

5. Place pallions of hard solder on the top piece (the flower), but do not exceed the marked line. The pallions need to be distributed evenly over the whole surface, not just the edges or the middle. This is to prevent creating an air pocket. If there is an air pocket, the air will expand when heated again and will create a blister.

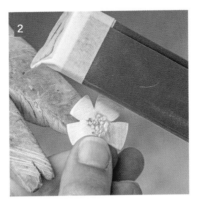

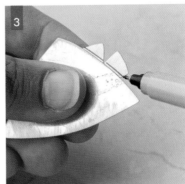

Using too little solder can cause an air pocket to form, creating a blister.

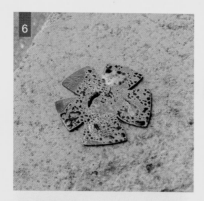

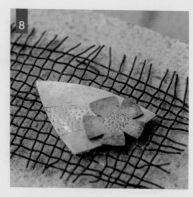

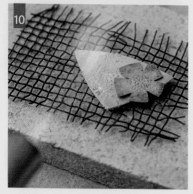

The triangle is heated ahead of the flower as it is bigger and is underneath the flower.

The triangle and the flower, sweat soldered together.

6. Heat to flow the solder over the surface. The solder does not need to cover the entire surface when it flows, as it will do this when you reflow the solder with the two pieces together.

7. Turn off the torch and set it safely aside. Position the top piece on the bottom one with the solder in between the two pieces. Because the bottom piece is usually bigger than the top one, the heat needs to be more concentrated on the bottom piece so that both pieces reach soldering temperature at the same time.

8. If you have steel mesh, place the piece on it. The steel mesh allows heat to go underneath the piece without lifting it.

9. If you do not have steel mesh, your piece will sit directly on the soldering block. Prop up the front edge on tweezers so that the flame goes under the sheet. When lifting the sheet with tweezers, lift the edge that is the shortest distance between the tweezers and the back of the sheet that is resting on the soldering block. This is to avoid the metal slumping like a hammock when heating as there is less metal and therefore weight unsupported by the soldering block. (See project 5, step 5b, page 133.)

10. Light the torch and heat the piece until the solder reflows. When the solder flows, you will see a silver line outlining the flower. Make sure that the solder flows all the way around the flower. Remember that solder runs to where the metal is hottest, so the solder may have run only on the side nearer to you where it might be hotter. Turn the piece around to check that the solder has run on all sides. If not, heat the piece to flow the solder where it has not flowed.

11. Pickle, rinse and dry.

TROUBLESHOOTING

If you have too little solder, the solder will not run all the way to the edges of the piece. You will see a black outline: a shadow between the two pieces.

Before you solder the two pieces together, if there are lots of areas with no solder, try spreading the solder around more with the back of the tweezers while it is flowing. If there are still lots of areas with no solder when the solder has run on the back of the top piece only, add some more solder with tweezers and reflow.

If you still have too little solder once the two pieces are soldered together, try reflowing the solder and pressing down on the top piece with tweezers to reduce the gap and force the solder to the edges. Do not push too hard, as it can mark the top piece.

If you need to add more pieces of solder, consider where you put the pallions. For this specific project, you can turn the brooch over and put the pallions on the flower against the triangle. If the flower were in the middle of the triangle, with no overhang, you would have to place the pallions around the outside of the flower, where the black line is, and reflow the solder. This method could leave solder marks on the front of the piece. To file these off, it is easier to put a slight curve or dome on the piece with your hands, so that when you file you are not filing lines across the whole piece.

Tap the piece flat again with a mallet when you have finished filing and sanding.

MAKING THE BROOCH FINDING

1. Using the remains of the sheet that you cut the triangle and flower from, mark a rectangle 3 × 8mm (⅛ × ⁵⁄₁₆in). You can use a ruler and scriber or dividers to mark it out.

2. Using a piercing saw, cut out the rectangle. This will become the brooch catch.

3. File the rectangle so it has straight edges.

4. Using a flat hand file or needle file, thin the top end of the rectangle so it is easier to bend. File about a third of the way down the rectangle, making a tapered wedge shape whereby the end is about half the thickness of the sheet.

5. Using a flat file, file the ends of the thin part into a curve.

6. Sand out all file marks.

7. Using round-nose pliers, bend the thinner end into a curve to make a hook.

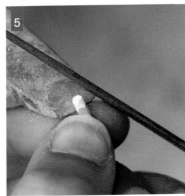

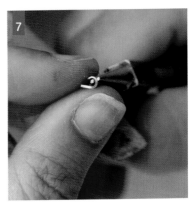

SOLDERING ON THE CATCH AND THE TUBE

Before you start soldering on the catch, consider carefully where the pin will go. Holding the brooch the right way up and looking at the back, the pin is always attached to the right-hand side. This is where the tube that holds the pin will go. Solder the tube above the middle line of the brooch so that it hangs well. If you solder it on the middle line, the brooch will tip forward when you wear it.

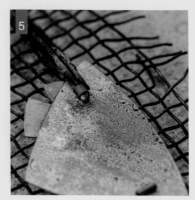

1. Place the piece upside-down either on the steel mesh on the soldering block or straight onto the soldering block, and flux the back.

2. Flux the tube and place it on the right-hand side of the brooch back and above the middle line. Place a pallion of easy solder against the side of the tube (not the end, to prevent the solder running into the tube).

3. Heat the brooch, avoiding the tube until the brooch is approaching soldering temperature, then move the heat across and around the tube until the solder flows.

4. Put your catch in reverse-action tweezers and flux the underside of it. Position the catch on the opposite side of the brooch to the tube, in line with the top of the tube. Rest the reverse-action tweezers down on the soldering block when it is positioned correctly.

5. Place a pallion of easy solder against the bottom of the catch. Solder in the same way as you did the tube. Remember to avoid reheating the tube too much.

6. Once soldered, pickle, rinse and dry.

HANDY HINT

Because you fluxed the whole back of the brooch, the catch can be soldered straight after the tube, without pickling in between. The flux will keep the back clean for soldering.

MAKING AND ATTACHING THE PIN

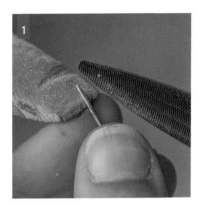

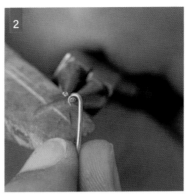

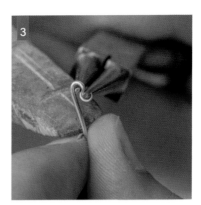

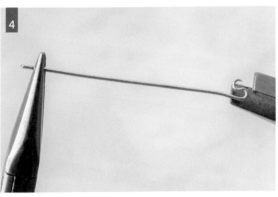

1. Take the 0.9mm (19-gauge) round wire and, with a flat file, file a taper on one end. (This does not need to come to a point, just a taper.) Sand out any file marks.

2. Using round-nose pliers, bend a curve on the end you have just tapered.

3. With the same round-nose pliers, adjust the wire in the nose of the pliers so you can hold the long end of the wire and bend it back around to get another curve, making an 'S' shape.

4. Hold the 'S' shape in flat-nose pliers and the other end of the wire in chain-nose pliers. Twist the flat-nose pliers, keeping the chain-nose pliers still, to harden the wire. Keep twisting until the hardness is enough to give the pin the spring it needs to work.

5. Thread the wire through the tube from bottom to top.

HANDY HINT

When you file a taper on a piece of wire, it helps to file a groove in your benchpeg and sit the wire in the groove while filing.

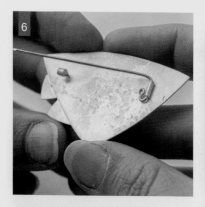

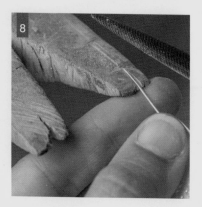

The finished brooch fittings.

6. When the wire pin is in as far as it will go, use your fingers to bend it at right angles to the tube. The pin also has to point upwards when the 'S' is resting on the brooch.

7. Put the pin in the catch. If the pin does not go in the catch, adjust the catch with round-nose pliers so that it does. Cut the pin with wire cutters just past where it comes out of the catch.

8. Now that the pin is the correct length, it needs to be filed, to a tapered point. To do this, file a groove in your peg and lay the wire in the groove to hold it while you are filing it. Turn it to get an even taper. Sand out the file marks.

HANDY HINT

If you go on to make brooches with bought findings and use a revolving safety catch, be careful when holding the catch in reverse-action tweezers while soldering. Hold the catch on the little tab that moves, not across the whole thing, otherwise it could get crushed with the squeeze of the reverse-action tweezers when heated.

FINISHING THE BROOCH

If there are any scratches on the flower, sand them off ready for polishing. Use one of these methods to finish your brooch:

BRASS BRUSH USED WITH A BURNISHER

Use a brass brush under running water with a drop of washing-up liquid (dishwashing liquid) to polish your brooch. Use a burnisher to highlight the edges.

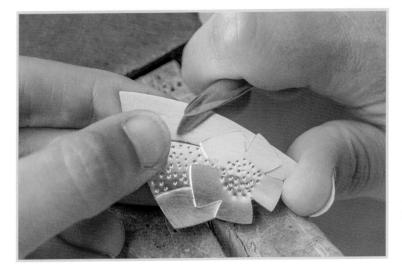

Use a burnisher to highlight the edges of your completed brooch.

BARREL POLISHER

The barrel polisher does not give a mirror finish on flat sheet, but you can still use it. To use a barrel polisher on cut card pieces, first lightly brass brush any areas where there are recesses that the shot won't reach: the shot will leave white halos around these areas (see page 56). You may, of course, like this effect, as it can enhance the piece. However, using the brass brush on those areas before barrelling means you get a more even finish after barrel-polishing.

REMOVING FIRESTAIN

Firestain is the copper (cupric oxide) in the sterling silver showing on the surface after heating. It looks like a grey-purplish shadow over the surface of the sheet. The finer you polish, the more it shows up.

Throughout the projects in this book, we have used polishing methods that generally do not show up the firestain. However, firestain shows up much more on flat, smooth silver sheet, such as in this cut card project.

If it is visible, it should be removed because it mars the appearance of the piece. As it is only surface deep, it can be removed. There are methods to help reduce firestain, such as fluxing the entire piece before soldering, but it is very hard to prevent altogether when using sterling silver.

Use one of these methods to remove firestain:

1. Sand the firestain off by using a sanding stick. If you can't reach the spot in this way, fold a piece of abrasive paper a couple of times to make a stiff corner to get into recesses where the firestain is. This is to prevent rounding off the edges of your piece or making undulated surfaces.

2. Use Tripoli on a felt stick and vigorously rub over the firestain.

3. Use depletion gilding – see below.

4. If you can find a Water of Ayr Stone (it is hard to find), this can be used to remove firestain. It is used by dipping in water and rubbing over the surface of the silver. It will remove the surface like a file but without file marks.

An example of firestain on silver sheet.

In the middle of the flower you can see that the silver is lighter where there is no firestain, but the rest of the flower has firestain all over it.

HANDY HINT

Putting tracing paper over the surface of the sheet can help you to see where the firestain is.

DEPLETION GILDING

This is a process of heating and pickling the piece several times. Heat the piece gently only to grey – it should not get pink or red. Put it in the pickle warm. Repeat until you do not see any grey when the flame is on the piece. Usually this needs to be repeated three to six times to get to this stage. The pickle dissolves the firestain (copper), leaving a thin layer of fine silver on the surface of the piece. There will be no firestain on the fine silver surface itself.

If using this method, finish with a polishing cloth or barrel polisher. The fine silver layer is very thin, so do not over-polish or you will go through it and show up the firestain again.

SAFETY NOTE

As depletion guilding involves putting the silver in the pickle when the silver is warm, remember to slide the piece between the lid and the bowl to avoid any splashback caused by the heat of the piece.

PROGRESSION

Make a cut card design where some of the edges line up.

Make a piece with mixed metals. Note the white halos around the top copper and brass pieces, left from polishing in the barrel polisher, as mentioned on pages 56 and 158. The piece on the left features brass, while the piece on the right features copper.

A stone setting has been added
to a cut card design, along with
texturing. The star has been cut out
with drill holes to make it look like
a falling star.

Make a cut card design with
a design cut into the top sheet.
A tiny star has been cut out of the
topmost star.

FUSED PENDANT
AND STUD EARRINGS
USING SILVER SCRAP

The finished fused pendant and stud earrings.

WHAT WE ARE MAKING

A pendant and stud earrings made from scraps of silver (from previous projects) that are melted together – or, fused.

You will always end up with scrap silver – although you can sell it back to a bullion dealer, it is more fun and economically viable to make jewellery with it.

CORE SKILLS

Sawing; forming; sanding; soldering; pickling; polishing.

NEW SKILLS

Fusing; melting silver into a ball; making stud earrings; making a bail for a pendant (alternative method); making a jump ring.

YOU WILL NEED

BENCH, HAND AND FORMING TOOLS

- Piercing saw and saw blades
- Beeswax or candle wax
- Needle files
- Round-nose pliers
- Wire cutters
- Flat-nose pliers
- Ruler
- Scriber

SOLDERING TOOLS

- Flux (such as borax) and dish or piece of slate
- Large soldering sheet
- Soldering block or charcoal block
- Paintbrush
- Torch for soldering
- Steel tweezers
- Reverse-action tweezers
- Tin snips
- Pickle in slow cooker
- Glass bowl with water and tongs (copper, brass or plastic)

FINISHING AND POLISHING TOOLS

- Abrasive paper
- Sanding stick (optional)
- Barrel polisher or brass brush
- Burnisher or steel teaspoon

MISCELLANEOUS ITEMS

- Plastic tea strainer or sieve (optional)
- Masking tape
- Marker pen
- Length of thread

MATERIALS

- Silver scrap from the other projects
- A length of 0.8mm (20-gauge) wire to make one or two jump rings for a bail
- From your length of 0.8mm (20-gauge) wire cut two 1.1cm (⁷⁄₁₆in) lengths for the posts of the earrings
- One pair of medium-weight scrolls for the stud earrings
- Easy solder

MAKING THE FUSED PENDANT

HANDY HINT

This is a good project to help you practise heating different-size pieces of metal. The bigger pieces need more heat to get to fusing temperature at the same time as the smaller pieces. If the smaller pieces get hotter first, they could melt into a ball.

1. Arrange some of the silver scrap you have left from the other projects on your soldering block in a shape that you like. You can cut bigger pieces into different shapes. The pieces must overlap, otherwise when they start to melt they will contract and have gaps between them and will not fuse together. There is no need to use flux.

HANDY HINT

Keep back a few pieces of scrap so that you can add them later if you need to adjust the shape or size of the design. You can keep adding pieces of scrap and fusing to the main piece without needing to pickle in between.

2. Heat the whole thing until it starts to shine and looks liquid. This is when they fuse together.

3. Quench and pickle. When you remove the piece from the pickle, you will see that the surface of the silver has a crinkled texture. This is called **reticulation**. Reticulation is used as an intentional texture.

4. If any of the edges are sharp, file them and sand where you filed.

ADDING MORE PIECES TO THE PENDANT

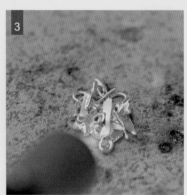

The piece is fused.

1. If you want to add more pieces to the pendant to improve its shape or appearance, place them on top of the already-fused pendant wherever you want them. You do not need to flux.

2. Remember, when you heat, the already-fused pendant is bigger than the new pieces, so it will need more heat to get to temperature. Aim to have all the pieces reach red at the same time.

3. After all the pieces are red, the piece will go shiny: this is the fusing process beginning.

4. If you carry on fusing, the pieces will start to look more rounded and unified and less like individual pieces. However, be careful to stop before you end up with one big ball.

Now, make a ball to solder onto the pendant – see overleaf.

HANDY HINT

You can stop heating the piece at any point to have a look at what you have got. You can then go back in again and heat areas that you think need more fusing.

MAKING THE BALL

Use a small piece of scrap silver. Imagine the scrap silver is like a piece of modelling clay that you roll between your fingers in order to visualize the size of the ball you will end up with.

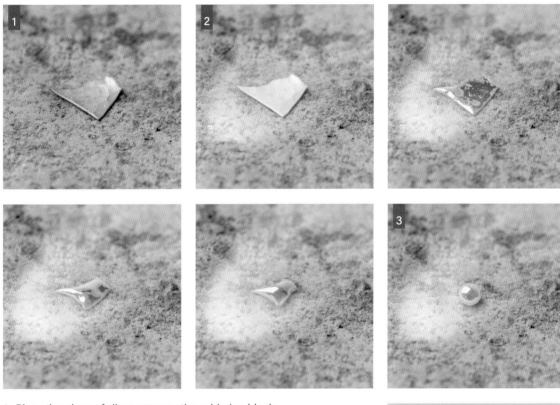

1. Place the piece of silver scrap on the soldering block.

2. Heat with the torch until it melts. This time you are melting the silver completely, so this is a chance to observe what the metal looks like as it builds up to melting point. Seeing the changes in colour and appearance shows you the stages that the metal goes through as you heat it. This will help you to know when you are overheating the silver during soldering.

3. The metal will mould into a ball with a flat bottom. The more metal you use to make a ball, the flatter the ball will become because of the effect of gravity.

4. You can add more pieces of scrap to make the ball bigger. If you want to do this, place the scrap so it is touching the ball and reheat until both melt together. It is easier to add pieces to make the ball bigger than removing parts when it is molten to make it smaller, so start small.

5. Quench the ball in water and put it aside where you will not lose it. You can pickle the ball, but it is not necessary because you can sand the bottom to clean it before soldering. This is not practical if you have multiple balls – see the Handy Hint on the opposite page.

HANDY HINT

If your ball is small, it can be difficult to get it out of the pickle with tweezers. Use a plastic tea strainer or small plastic sieve to suspend the ball in the pickle – this will make it easy to get out. If you make multiple balls, you can get them all out of the pickle at the same time using the same method.

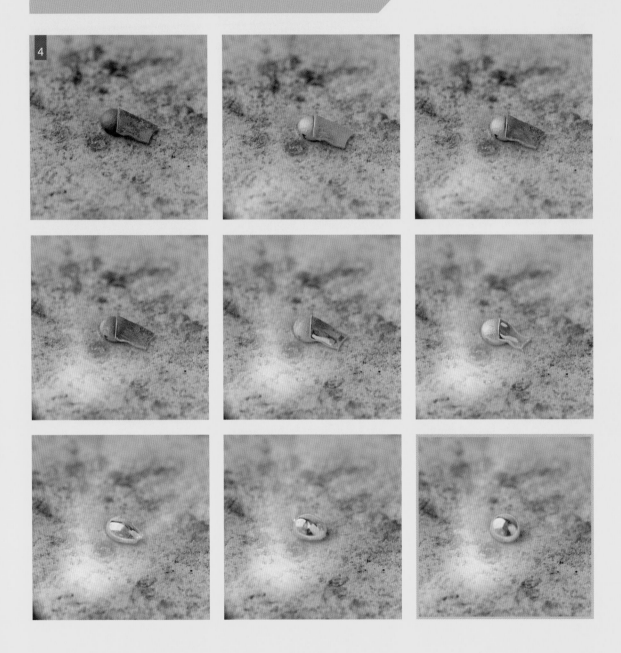

SOLDERING THE BALL ONTO THE PENDANT

1. Place the pendant on the soldering block.

2. Place a piece of hard solder on the soldering block away from the pendant.

3. Flux the spot on your pendant where you have decided the decorative ball will go.

4. Flux the flat, clean side of the ball and put it on the solder.

5. Heat to melt the solder onto the ball.

6. Pick up the ball with tweezers and place it on the fluxed spot so the solder is between the pendant and the ball.

7. Reflow the solder by heating the pendant. You will see a silver line when the solder flows at the bottom of the ball. This is sweat soldering, as in project 6.

8. Quench and pickle.

HANDY HINT

Remember to hold the torch in your non-dominant hand and the tweezers in your dominant hand, so that if the ball does move when soldering you can move it back with the tweezers. It is amazing how easy it is to forget that the piece is very hot, and when something moves you go to stop it with your hands. To avoid this, have your tweezers at the ready.

ADDING A BAIL

If your design has a convenient hole in it through which you can put a jump ring for a bail, use it.

1. Make a jump ring big enough to go through the hole and for a chain to go through it. (See page 88 for instructions on making one jump ring.)

2. Place the jump ring through the hole and put a piece of easy solder on the join.

3. Solder the jump ring closed (as in project 4, page 119).

4. Quench and pickle.

ATTACHING THE BAIL IF THERE IS NO HOLE

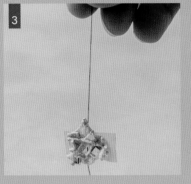

HANDY HINT

If you are soldering two pieces together and one is quite large, the solder will stay molten longer. Let it cool a little to avoid moving it while the solder is still molten, otherwise pieces may move.

If there is no convenient hole for the jump ring, follow these steps.

1. Make two jump rings from your length of 0.8mm (20-gauge) round wire, one smaller than the other, using round-nose pliers. Form the smaller one on the narrow end of the pliers and form the larger one around the widest part of the pliers. Make them one at a time. (See page 88.)

2. Close the join on the small jump ring ready for soldering.

3. To work out where to place the bail, stick the end of a piece of thread down on the back of the pendant with masking tape. Hold it up on the thread and see if it hangs as you would like. Move the thread until it is in a spot where the pendant hangs nicely. Mark that spot with a marker on the front.

4. Flux all areas to be soldered: the jump ring join and the back of the pendant where the bail will go.

5. Place the pendant face-up and the small jump ring on the soldering block away from each other.

6. Put a largish pallion of easy solder on the join of the small jump ring and solder. This solder will be reflowed to join it to the back of the pendant, which is why the pallion of solder needs to be a bit bigger than the normal size used just to solder a jump ring closed.

7. Place the pendant on top of the small jump ring at your mark so half of the jump ring is showing above the top of the pendant. Position the two pieces on the soldering block so you can see the jump ring from the side.

8. Heat to reflow the solder that is on the jump ring. The two pieces need to reach soldering temperature at the same time, so concentrate the heat on the pendant first and move across to the jump ring. You will see a silver line at the edge of the jump ring and pendant when it flows.

9. Quench in water.

10. Put the larger jump ring through the smaller jump ring and align the join for soldering.

11. Place the pendant down on the soldering block. The large jump ring will sit up. Position with the solder join at the top. Place a small pallion of easy solder on the join. Solder.

12. Now the jump ring is soldered, it is ready to be pickled, rinsed and polished.

FINISHING THE PENDANT

Polish the pendant with either the barrel polisher or the brass brush. Remember to brass brush areas that will stay white where the balls of the barrel polisher do not reach if you do not want those areas to stay white.

High spots can be accented with a burnisher.

MAKING FUSED STUD EARRINGS

Two similar collections of scrap prior to fusing.

The fused silver scrap earrings.

1. Using the same methods as for the pendant, make two fused pieces to turn into earrings. They can be whatever size you want, depending on how much scrap you have. You can aim for mismatched earrings or you can try to make them similar.

2. Once fused, file any sharp edges with a file. Sand those areas if necessary.

3. Take the 1.1cm (7⁄16in) lengths of 0.8mm (20-gauge) wire and file one end of each flat. These will be the earring posts.

4. Flux the flat ends ready for soldering.

5. Place both of the earrings face-down on the soldering block, separate from each other, and flux the backs where the posts are going to go.

6. Pick up one of the 1.1cm (7⁄16in) lengths of wire in reverse-action tweezers and place it with the flat end pointing downwards, where you want it on the back of the earring. Rest the reverse-action tweezers on the soldering block in that position. Add a pallion of easy solder on the back of the earring against the post.

7. Solder, taking care not to overheat the post, by heating the body of the earring first. Remember to let the earring cool a bit before you pick it up to ensure that the solder is definitely not molten.

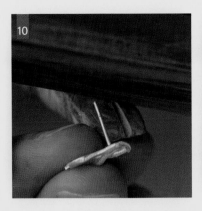

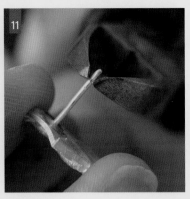

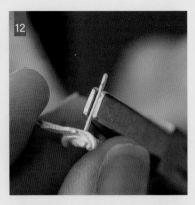

8. Repeat steps 6 and 7 with the other earring.

9. Pickle, rinse and dry both earrings.

10. File or sand the end of the earring posts so that there are no sharp points.

11. Using wire cutters, gently grip the end of the post approximately 1mm (¹⁄₁₆in) from the end you just sanded. Twist the post around in the wire cutters. Do not do this too roughly, or you will cut off the end of the post. This process will make a small groove in the post that will help to secure the scroll back.

12. Using flat-nose pliers, adjust the posts so that they are straight.

13. Polish the earrings with either a barrel polisher or brass brush.

14. Finally, add the scroll backs.

HANDY HINT

To increase the strength of the solder join of the posts to the earrings, you can part drill a hole to make a recess for the post to sit in. Use the same drill bit that you used in project 4 (see pages 113 and 117). It does not need to be the same diameter as the posts. This recess increases the surface area of the solder join, making it stronger.

ADJUSTING SCROLL BACKS

Scroll backs are often too tight when they are new. To loosen, push the nose of a pair of round-nose pliers into the curls of the scroll and gently pull them apart. Test the fit on the earring post. If it is now too loose, squeeze the curls together with your fingers from the sides.

PROGRESSION

Multiple balls have been made and jump rings have been soldered onto them. A larger ring has been made out of thicker round wire and threaded through the jump rings on the balls and then soldered shut. Another smaller jump ring is soldered to the top of the larger ring as a bail.

You could also choose to oxidize some of the balls and add them to jewellery pieces as decorative elements.

Balls have been made and soldered to a rectangular wire to make a pendant. Gold behaves in exactly the same way as silver; here, a red-gold ball has been soldered on as a design interest.

Combine fused silver balls with previous projects. Here, small balls have been made and soldered onto rings. These rings would make perfect stacking rings if they were made the same size.

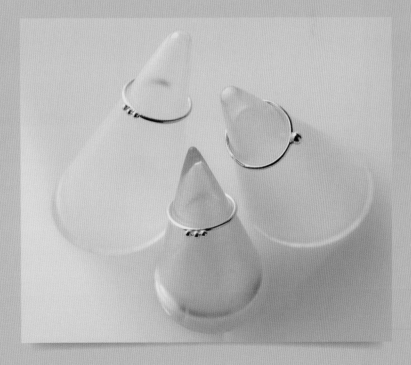

This piece features multiple fused pieces. Three have been linked together with jump rings (see Making Jump Rings, pages 88–91), and then attached to a ready-made chain, to make a statement necklace.

Fused pieces have also been used to make the toggle clasp. A 'V' punch has been used to add an accent on the clasp.

Using the same method that you used in the project to solder earring posts, solder a tie tack to the back of a piece you have made to make a tiepin.

These cufflinks are made from four fused pieces. The two larger pieces form the front and the two smaller pieces go through the buttonhole. Each cufflink is linked with an oval ring and two half rings made of 1mm (1⁄16in) round wire.

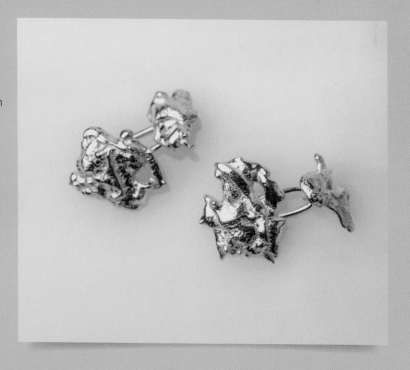

MOVING FORWARDS

Now that you have completed the seven projects, you have been introduced to enough techniques to move forwards and use what you have learned to start making your own designs. This is the time to consolidate your experience through practice, becoming more confident every time you have a go.

WHAT SHALL I MAKE?

Designing, drawing and technical skills need to be developed alongside each other – they are processes that help to underpin creative ideas. It might be reassuring to know that a sudden flash of inspiration is a rare event, and may never happen: almost nobody sits down with a blank piece of paper and comes up with a perfect design. To demystify the idea of design, think of it as answering a series of questions:

- Do I want to make a pendant or earrings?
- If I make earrings, will they be studs or dangles?
- What size will work? (Remember to consider the weight of the finished piece as well.)
- Will there be a texture?
- Is a stone setting part of the design?

These decisions are the starting points to help you develop a plan.

HANDY HINT

Resist the urge to include photographs of jewellery in your scrapbook as you may inadvertently copy other people's work rather than developing ideas of your own (see Considering Copyright *on the opposite page).*

HOW WILL IT LOOK?

Designing is a skill that can be nurtured and enhanced. Creating a scrapbook is a positive first step that encourages you to pull together ideas that you like. Your scrapbook then becomes a mood board or image library that you can dip into for inspiration. It could feature solely drawings, or you could add your own photographs and pictures from various sources.

HOW WILL I MAKE IT?

Initially, pick a technique or two from the projects and develop a design of your own that incorporates them. You can use the Progression sections to help you decide what to make.

CONSIDERING COPYRIGHT

While all the projects and progression suggestions in this book are intended for you to copy and use as you wish, be aware that the law gives people and businesses property rights to the information and intellectual goods they create in the form of copyrights, patents and other protections. This means that photographs, jewellery designs (drawn or existing), artworks and so on are legally protected from being copied.

'It is not always clear cut where the line is between inspiration and imitation, but generally any design based on an existing original design will be likely to copy the "important elements" of that design if it is used for inspiration.'

– Jane Banyai,
ACID (Anti Copying in Design, UK)

In other words, to steer clear of any issues, try to base your designs on your own resources. Take your own photographs and make your own doodles and drawings. If you are using existing photographs from magazines or books, develop your ideas as in the examples in this chapter, avoiding using a shape or pattern exactly as shown in the original image.

If you would like further information, look at the websites of organizations such as ACID in the UK.

HANDY HINT

Within the projects, you will have been told the order in which to do things. For your own designs, you will have to work this out for yourself using what you have learned.

THE FOUR-STEP PLAN

When designing and making a piece, you may wish to follow this four-step plan to help you see your designs through to fruition.

1. Research;

2. Development;

3. Fabrication plan;

4. Analysis.

1. RESEARCH

Here are some different ways you can build up a collection of images that will help you to start designing.

- Sketch shapes that appeal to you – these can be anything from quick doodles to elaborate drawings.

- Cut out images from magazines, flyers and so on that catch your eye, focusing on forms, patterns, textures and shapes.

- Take photographs of particular things that you like – maybe flowers, wrought-iron work, architecture or items around the house.

- Go to museums, galleries, shops or wherever you feel stimulated visually, and be conscious of the images around you as you spend time somewhere you enjoy.

- Carry a sketchbook or small pad everywhere and jot down ideas that you have or images that you see when you are out and about.

Shape, form and detail all around us can inspire ideas.

From these freehand scribbles, interesting shapes have been highlighted to see if potential designs can come from them.

2. DEVELOPMENT

Once you have built up a sketchbook, you can use it to develop designs. Remember that your first attempts may lack finesse, but persevere and they will develop into something you like.

Here are some ideas:

- Put tracing paper over an image and trace a section or sections. To make a symmetrical shape (such as the basis for a pendant), fold the tracing in half and trace the shape again.

- Cut a small hole approximately 4cm (1½in) wide in a piece of paper – it can be square, oval or round: it is up to you. Place the piece of paper over one of your images so that it frames one section. Move it around and you will see the image in a different way, isolating shapes and forms.

Here, ideas are in progress from the starting point of the doodles on the opposite page. You can see where the tracing paper has been folded in half to trace a symmetrical design.

Here, a collage has been made from magazine images and a piece of paper with a hole is being used to isolate shapes and forms. Elements of the design viewed within the hole have been isolated and redrawn, above.

- Draw a variety of quick sketches based on the images you have isolated with the hole in the paper. Put away the original pictures and work from your sketches only. This will help you to develop your idea without being influenced by the original picture. Pick out the parts you like and use these as a basis as you develop your design.

Once you have a design, you need to decide on its size. Draw your design at the actual size at which you want to make it. To test that you have chosen the right size, put your drawing against the body to see how it looks when worn.

Two size options from the doodles on the previous pages have been drawn onto tracing paper side by side, to help decide the most preferable size.

- If you are computer-minded, you can take a photograph of a page in your scrapbook or of one of your sketches; open the picture in an image-editing program (such as Adobe Photoshop) and manipulate it to see what you get. You can change the proportions, make it black and white, reverse the colours, crop it or even warp it.

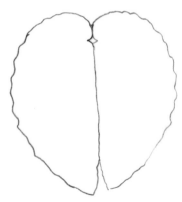

A tracing is made of the shell in the photograph. The tracing is then scanned so that it can be copied, then enlarged or made smaller.

Here, the shapes have been distorted by making them taller or wider to see if that improves the shape. They have also been repeated to see what effect that gives. A circle has been drawn around a small version of the drawn shell to see if it would work well with cut card work.

Once the size has been chosen, trace off the separate component parts that will make up your piece. Now you are ready to make your fabrication plan.

     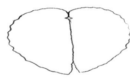

THE EVOLUTION OF AN IDEA

Many of the techniques used on pages 178–180 have been used in the worksheet below. The photograph of the flower has been digitized to allow manipulation of the image. It has been made simpler, focusing on the outline and colour. The size has been adapted for various pieces of jewellery, such as earrings, a necklace, a bracelet and a ring. Different sizes have been used in combination on the necklace to see how they go together. Possibilities for different coloured stones are illustrated to fit the colour scheme; options for the findings and connections have also been drawn.

Together, all the details on this worksheet can lead to the making of a whole suite of jewellery.

3. FABRICATION PLAN

Now that you have a design, you need a fabrication plan. This is how you are going to make your piece and in what order. The sequence is important because certain things will need to be done before others.

- Decide how many separate elements there are. Ask yourself what the component parts are, for example a back sheet, stone setting, bail or jump ring.

- It may help to label your design so that you have a list of all the component parts and what materials you will need to make them.

- Work out which techniques you will use in making your piece, for example texturing, cutting out a shape or making a stone setting. These processes should follow a logical sequence.

These earrings have been created based on the sketches on pages 178–180.

Consider these points:

▶ Having your component parts on separate pieces of tracing paper will enable you to move them around to work out the most economical arrangement on your piece of silver sheet before you stick them down.

▶ If there is a texture, when is it best to apply it? For example, it is often not practical to add texture after you have domed a piece, so texture before doming.

▶ The order in which you solder pieces together is important. For example, with a cut card project you would need to sweat solder the sheet together before adding any findings.

▶ Remember, always start with hard solder. If you have elements that are soldered separately from the main piece, use hard solder for all of them, and then use medium or easy solder to bring together all the parts.

▶ Stones need to be set after all soldering is finished, because most are damaged by the extreme heat of soldering and/or the pickling chemicals (they could crack, discolour or cloud).

Once you have your fabrication plan, make your piece. You may need to adjust as you go along – the process of making will determine whether your fabrication plan was accurate.

4. ANALYSIS

Once you have finished, analyze what you have done. Think about whether you are happy with the design or if there are elements that could have been changed to make it better. Would fabrication have been easier by swapping to thicker or thinner metal or by rejigging how you put the piece together?

In future projects, repeat the things that went well and improve on the things that didn't. This could be the order of soldering or the final look of the design. Remember, you can't move forwards without making mistakes – they will undoubtedly happen. Don't despair – look upon them as constructive experiences that provide an opportunity to learn. The process of analysis improves your technical abilities and refines your designs. Each piece that you make will teach you something to take forwards to your next project.

HANDY HINT

Think WWW: What Went Well.

Consider all the things that went well in the project. It is equally important to focus on the things that went as planned, so that you can repeat those in the future, rather than only remembering the things that went wrong.

Each item made is a learning experience, so give yourself a pat on the back for the things that went well.

3) Fabrication Plan

Earrings

a. Cut out the 6 shapes from sheet.

b. Make 6 jump rings to join hanging detail to the studs section.

c. Decide which way the 'wings' are to be soldered, which one on top.

d. Decide if the top one needs texture.

e. Texture now if going to.

f. Solder together the 'wings' stud parts x 2.

g. Solder the jump rings to all 4 units (studs and dangle detail).

h. Cut posts x 2.

i. Half drill studs where posts are to be soldered.

j. Solder on posts.

k. Join 2 units together (stud and dangle unit x 2) with jump rings.

l. Solder jump rings.

m. Polish.

4) Analysis

a. Don't make all jump rings together: wait until you know what's definitely going to hang below.

b. Don't cut corners – hammer lines in on steel block, because the lines are messy when hammered on wood.

c. Make again so the lines are not messy and maybe add more.

d. Change the shape of wings part as it has lost something from the drawing.

Janet's sketchbook complete with her notes on her Fabrication Plan and Analysis.

GLOSSARY

Alloy:

A mixture of metals. Sterling silver is a mixture of silver and copper.

Annealing:

The process of heating, softening and cooling a metal to make it malleable (i.e. easier to twist, bend, hammer and press without it fracturing). See also: work hardening.

Bail:

The part of a pendant that a chain goes through.

Bezel:

A band of metal that goes around a stone to set it in place.

Borax:

A white mineral that is used dissolved in water as a flux for use in soldering.

Burnish:

To polish by rubbing to compress with a polished steel tool such as a burnisher or other steel item such as the back of the bowl of a steel teaspoon.

Burr:

A rough edge or ridge of metal created when filing or sawing.

Cabochon:

A gemstone with a flat bottom and a domed, polished top that is not faceted.

Capillary action:

The ability of liquid to flow into a narrow space irrespective of gravity.

Cut:

Used to describe the coarsenesses of hand files and needle files. The cut is described by numbers: the higher the number, the finer the file is. A fine-cut file removes less material than a medium- or coarse-cut file.

Findings:

The functional parts of jewellery, such as jump rings, catches, earring hooks, posts and clips.

Fine silver:

Silver that is not alloyed, so it is 99.9% silver, with the balance being trace amounts of impurities. This is different from sterling silver, which is 92.5% silver.

Firestain:

A deposit of cupric oxide that forms on sterling silver when it is heated. It shows as a purplish stain on the surface of the metal when it is polished.

Flux:

A substance put on the join of a piece to prevent oxides from forming on the join during soldering. When put on the join, the flux keeps it clean during soldering so that the solder can flow.

Former:

Anything used to form metal around to create a desired shape. It can be a purpose-made tool such as a ring mandrel, or anything that is the right shape and hardness to form around, such as wooden dowelling, chopsticks, nails, pens and so on.

Gauge:

A standard unit of measurement for the thickness of silver sheet or diameter of wire.

Pallion:

The name given to pieces of solder once they have been cut. These can also be called 'panels.'

Patina, patination:

Patination is the process of changing the surface colour of metal through exposure to acids, chemicals, heat or air. The resultant colour is called a patina.

Pickle:

An acid solution used to remove flux and oxides from metal once it has been soldered.

Quench:

To cool hot metal rapidly, usually in cold water.

Reticulation:

The crinkled surface on sterling silver when it has been heated enough to melt the top layer of metal.

Ring shank:

The part of the ring through which the finger goes.

Rouge:

A red polishing compound usually used as the final stage for a mirror finish.

Solder:

Silver solders are different silver alloys that melt at lower temperatures than pure silver so parts can be put together during the soldering process without the main piece melting.

Soldering:

The process used to join together the different components of jewellery by melting a lower melting temperature alloy between them. (Technically, the process is brazing, because the temperatures involved are above 430°C, or 806°F.)

Tripoli:

A brownish polishing compound that is rougher than rouge and used to remove scratches and small marks. Used before rouge.

Work hardening:

The process in which metals get harder as they are bent or hammered. The metal often gets to a stage where it is too hard to work with and then it is annealed to make it malleable (i.e. workable) again.

A WORD ABOUT HALLMARKING

We hope that this book will lead you to many enjoyable years of jewellery making. You will find that people will comment on your jewellery and maybe ask you to make pieces for them. If you do get to the point of selling jewellery, you need to be aware of the hallmarking laws in the country where your pieces are on sale.

In the UK, any precious metal items over a certain weight have to be hallmarked by law. Hallmarks guarantee the authenticity and content of precious metals, thereby protecting the purchaser. A hallmark consists of a minimum of three marks:

• sponsor's (or maker's) mark;

• assay office mark;

• millesimal fineness mark: this denotes what material the piece is made from.

For example, 925 represents sterling silver – an alloy that is 925 parts silver out of 1,000, combined with 75 parts other metals such as copper. If you see the 925 stamp by itself, without the other two marks, the metal has not been tested and verified by an official assay office.

For more information about hallmarking, visit the following websites (UK only):

London: www.assayofficelondon.co.uk

Edinburgh: www.edinburghassayoffice.co.uk

Sheffield: www.assayoffice.co.uk

Birmingham: www.theassayoffice.com

APPENDICES

TABLE OF RING SIZE CONVERSIONS

NOTE

This chart gives a good indication of how different ring sizes relate to each other, but, as is often the case with different measuring systems, there is not always an exact match.

USA/Canada	Australia/Ireland/New Zealand/UK	Europe	India/China/Japan
½	A	38	
¾	A½	38	
1	B	39	1
1¼	B½		
1½	C	40.5	
1¾	C½		
2	D	42.5	2
2¼	D½	42.5	
2½	E	43	3
2¾	E½	43	
3	F	44	4
3¼	F½		5
3½	G	45	
3¾	G½		6
4	H	46.5	7
4¼	H½	46.5	
4½	I	48	8
4¾	J	49	
5	J½		9
5¼	K	50	
5½	K½		10
5¾	L	51.5	
6	L½		11
6¼	M	53	12
6½	M½	53	13

USA/Canada	Australia/Ireland/New Zealand/UK	Europe	India/China/Japan
6¾	N	54	
7	N½		14
7¼	O	55.5	
7½	O½		15
7¾	P	56.5	
8	P½		16
8¼	Q	58	
8½	Q½	58	17
8¾	R	59	
9	R½	59	18
9¼	S	60	
9½	S½		19
9¾	T	61	
10	T½		20
10¼	U	62.5	21
10½	U½		22
10¾	V	64	
11	V½	64	23
11¼	W	65	
11½	W½		24
11¾	X	66	
12	X½		25
12¼	Y	68	
12½	Z	69	26

PIERCING SAW BLADE SIZES

8   7   6   5   4   3   2   1   1/0   2/0   3/0   4/0   5/0   6/0   7/0   8/0

← *coarsest*   *finest* →

Sizes 2/0, 3/0 and 4/0 are most commonly used for jewellery-making purposes. The important thing is that there is more than one tooth for the thickness of metal being cut.

TABLE OF GAUGES AND SIZES

These charts show conversions between different measuring systems for wire gauge. Where there is no exact equivalent, we have used an approximation. American Wire Gauge (AWG), also known as the Brown & Sharp (B&S) wire gauge, is used in North America to measure wire and sheet thickness. There are other, different gauge systems that can be applied, such as the British Standard Wire Gauge (SWG), but we have not listed that here as the metric system is used in Europe.

Metric (mm)	Gauge (B&S)	Imperial (inches)	Imperial (inches as fractions)
0.3	28	0.012	
0.4	26	0.016	$\frac{1}{64}$
0.5	24	0.020	
0.6	23	0.024	
0.7	21	0.028	
0.8	20	0.031	$\frac{1}{32}$
0.9	19	0.035	
1.0	18	0.039	
1.2	17	0.047	$\frac{3}{64}$
1.5	15	0.059	
1.6	14	0.063	$\frac{1}{16}$
2.0	12	0.078	$\frac{5}{64}$
2.4	11	0.094	$\frac{3}{32}$
2.5	10	0.098	
2.8	9	0.109	$\frac{7}{64}$
3.0	9	0.118	
3.2	8	0.125	$\frac{1}{8}$
3.6	7	0.141	$\frac{9}{64}$
4.0	6	0.156	$\frac{5}{32}$
4.4	5	0.172	$\frac{11}{64}$
4.8	5	0.188	$\frac{3}{16}$
5.2	4	0.203	$\frac{13}{64}$
5.6	3	0.219	$\frac{7}{32}$
6.0	3	0.234	$\frac{15}{64}$

Gauge (B&S)	Metric (mm)	Imperial (inches)	Imperial (inches as fractions)
29	0.29	0.011	
28	0.32	0.013	
27	0.36	0.014	
26	0.41	0.016	¹⁄₆₄
25	0.46	0.018	
24	0.51	0.020	
23	0.57	0.023	
22	0.64	0.025	
21	0.72	0.028	
20	0.81	0.032	¹⁄₃₂
19	0.91	0.036	
18	1.02	0.040	
17	1.15	0.045	³⁄₆₄
16	1.29	0.051	
15	1.45	0.057	
14	1.63	0.064	¹⁄₁₆
13	1.83	0.072	
12	2.05	0.081	⁵⁄₆₄
11	2.31	0.091	³⁄₃₂
10	2.59	0.102	
9	2.91	0.114	
8	3.26	0.129	¹⁄₈
7	3.67	0.144	⁹⁄₆₄
6	4.12	0.162	⁵⁄₃₂
5	4.62	0.182	³⁄₁₆
4	5.19	0.204	¹³⁄₆₄
3	5.83	0.229	⁷⁄₃₂

MOHS SCALE OF HARDNESS

Stones are ranked on the Mohs scale of hardness, where 10 is the hardest and 1 is the softest.

ALPHABETICAL

Gemstone	Hardness
Agate	7
Alexandrite	8.5
Amber	2.5
Amethyst	7
Ametrine	7
Apatite	5
Aquamarine	7.5
Aventurine	7
Beryl	7.5
Bloodstone	7
Carnelian	7
Chalcedony	7
Citrine	7
Cubic zirconia	8.5
Diamond	10
Emerald	7.5
Feldspar	6
Fluorite	4
Garnet	6.5
Jade	6
Jasper	7
Jet	2.5
Labradorite	6
Lapis lazuli	5.5
Lolite	7
Malachite	4
Moonstone	6

Gemstone	Hardness
Morganite	7.5
Obsidian	5
Onyx	7
Opal	6
Pearl	3
Peridot	6.5
Quartz	7
Red beryl	7.5
Ruby	9
Sapphire	9
Spinel	8
Sunstone	6
Talc	1
Tanzanite	6.5
Tiger's eye	7
Topaz	8
Tourmaline	7.5
Tsavorite	6.5
Turquoise	5
Zircon	7.5

BY HARDNESS

Gemstone	Hardness
Diamond	10
Ruby	9
Sapphire	9
Alexandrite	8.5
Cubic zirconia	8.5
Spinel	8
Topaz	8
Aquamarine	7.5
Beryl	7.5
Emerald	7.5
Morganite	7.5
Red beryl	7.5
Tourmaline	7.5
Zircon	7.5
Agate	7
Amethyst	7
Ametrine	7
Aventurine	7
Bloodstone	7
Carnelian	7
Chalcedony	7
Citrine	7
Jasper	7
Lolite	7
Onyx	7
Quartz	7
Tiger's eye	7

Gemstone	Hardness
Garnet	6.5
Peridot	6.5
Tanzanite	6.5
Tsavorite	6.5
Feldspar	6
Jade	6
Labradorite	6
Moonstone	6
Opal	6
Sunstone	6
Lapis lazuli	5.5
Apatite	5
Obsidian	5
Turquoise	5
Fluorite	4
Malachite	4
Pearl	3
Amber	2.5
Jet	2.5
Talc	1

INDEX

alloy(s)   41, 51, 184, 185
annealing, work hardening   51, 77, 78, 80, 81, 106, 120, 184, 185
anvil   11, 30, 61, 117
    attaching   30
    G-clamp   11, 30

bail   23, 113, 117, 119, 121, 163, 168, 169, 172, 182, 184
benchpeg   10–11, 29, 30, 34, 35, 36, 38, 40, 61, 63, 64, 67, 88, 89, 117, 129, 134, 156, and every project
    attaching to the anvil   30
    sawing   30
bezel(s)   23, 26–27, 122–125, 126–145, 184
    getting the correct height   124–125
    shaping around irregular stones   144
    soldering   129, 131, 132, 135, 140, 142, 145
    trimming and filing the sheet   134
    working out the length   23, 123
bezel strip   23, 26–27, 122, 123, 124–125, 126–145
    pros and cons   122
borax, see also: flux   18–19, 45, 49, 50, 61, 77, 93, 113, 127, 147, 163, 184
brooch(es)   17, 23, 27, 46, 104, 146–161 ,
    brooch design(s)   23, 148, 150
        cutting out   150
        preparing   148
        transferring onto metal   149
    finishing   158
    texturing   151
burnishing   20–21, 54, 56–57, 61, 73, 110, 113, 122, 127, 136, 137, 139, 141, 146–161, 163, 170, 184
    highlighting by   21, 57, 73, 137, 141, 158
burr(s)   64, 82, 88, 91, 116, 117, 150, 184,

capillary action   51, 82, 184
cut card technique   17, 23, 46, 146–161

doming block(s)   13, 106, 113
doming, punch(es)   12–13, 90, 106, 113, 118
    choosing your doming punch   118

earring(s)   27, 46, 47, 64, 89, 92–103, 120, 142, 162–175
    fused stud   162–175
    headpin(s)   93, 98, 100
    hook(s)   93, 94, 98–99, 101, 142, 184
    wire wrap loop(s)   24–25, 93, 99, 100, 103, 142

file(s)
    care of   40
    fitting a handle to a hand file   39
    use of   40
filings   27, 39, 40
findings   147, 154–157, 181, 182, 184
firestain   122, 159, 184
    removing   159
        depletion gilding   159

flux, fluxing   18–19, 29, 45, 46, 49, 50, 51, 52, 61, 69, 77, 93, 95, 96, 97, 98, 113, 127, 130, 131, 132, 133, 135, 147, 152, 155, 159, 163, 164, 165, 168, 169, 170, 184
former   90, 91, 99, 103, 107, 184

gauge(s)   25, 27, 61, 77, 93, 94, 98, 104, 113, 119, 122, 127, 129, 131, 132, 136, 145, 147, 151, 156, 163, 169, 170, 184, 188–189
    gauges and sizes   188–189

hallmarking   185
health and safety   17, 19, 29, 38, 48, 52, 56, 111, 131, 159

jump ring(s)   17, 25, 42, 88–91, 93, 94, 95, 97, 102–103, 113, 117, 119–120, 163, 168–169, 172, 174, 182–184
    closing   94
    drilling a hole for   117

Mohs scale of hardness   40, 99, 139–141, 143, 190–191

oxidizing   14–15, 29, 45, 110–111, 119, 121, 172,

patina, patination   111, 184
pendant(s)   23, 27, 42, 64, 75, 104, 112–121, 142, 145, 162–175, 176, 179, 184
    cutting out the circle   115
    doming   118
    finishing   119, 170
    fused   27, 64, 162–175
    marking out a circle   114
pickle, pickling   17, 19, 29, 47, 49, 50, 51, 52–53, 61, 68, 69, 77, 78, 82, 93, 95, 97, 98, 113, 119, 127, 129, 131, 135, 147, 153, 155, 159, 163, 164, 166, 167, 168, 169, 171, 182, 184
piercing saw   10–11, 14, 30, 34, 36, 61, 64, 77, 93, 110, 113, 115, 121, 127, 129, 134, 139, 147, 149, 150, 154, 163, 187
    blade(s)   14, 34–35, 36, 37, 38, 61, 64, 77, 91, 93, 113, 127, 129, 132, 134, 139, 147, 163, 187
        sizes   187
    frame(s)   14, 34–35, 37, 38
    setting up   34–35

quenching   51–52, 69, 97, 131, 164, 166, 168, 169, 185

reticulation   164, 185
ring(s)   10–17, 23, 25, 38, 53, 57, 60–87, 103, 109, 120, 126–145, 173, 181, 184, 185, 186–187
    back sheet   129, 132–133, 135, 139, 142–143, 145, 182
    round-wire   62, 65, 69, 72, 82, 83
    shank   135–136, 142, 185
    size conversions   186–187
    sizing   11, 61, 62, 64, 77, 81–82
    starting   65–66
        bending the wire   65–66
        making a good join   65–66

stone setting for   47, 122–125, 126–145
    square-wire   57, 62, 65, 66, 68, 69, 71, 72, 73, 81, 82, 83, 84

sanding   21–23, 45, 51, 53, 54–55, 61, 71–73, 77, 84, 93, 109, 111, 113, 115, 116, 117, 118, 121, 125, 127–129, 130, 131, 134, 135, 140, 147, 150, 152, 154, 156–159, 163–164, 166, 170, 171
satinizing   22, 109, 110
sawing   10–11, 30, 36–38, 45, 61, 63–64, 67, 77, 88–89, 91, 93, 113, 115, 127, 147, 150, 163, 184
    corners   37, 150
    curves   37
    inside shapes   37
    straight lines   36
silver scrap   25, 27, 39, 64, 129, 131, 162–175
silver sheet   23, 26–27, 104, 112–121, 122, 127, 146–161, 182, 184
solder, soldering   16–17, 19, 24–25, 29, 41–51, 61, 63, 66–67, 68–72, 74, 77, 78, 82–83, 87, 89, 93, 94, 95–97, 98, 103, 107, 113, 119, 121, 125, 127, 129–133, 135, 136, 139, 140, 142, 144–145, 147, 151, 152–155, 157, 159, 163, 164, 165, 166, 168–171, 172, 173, 175, 182–185
    easy   25, 41, 42, 93, 97, 113, 119, 127, 135, 147, 155, 163, 168, 169, 170, 182
    flowing   42, 46, 48–51, 82, 131, 133, 152, 154,
    hard   25, 41, 42, 61, 69, 77, 82, 93, 95, 127, 129, 147, 152, 168, 182
    join(s)   42, 69, 71, 72, 74, 82, 83, 89, 121, 130–131, 135, 140, 144, 169, 171
    medium   25, 41, 42, 43, 127, 131, 132, 133, 182
    pallion(s), panels   19, 43, 46, 95, 96, 97, 129, 131. 135, 152, 154, 155, 169, 170, 184
    strip   41–43
soldering surface(s)   19, 43, 47, 152
    choosing   43
    supporting your work   47
soldering torch   17, 29, 47–48, 61, 69, 77, 93, 96, 98, 113, 127, 147, 153, 163, 166, 168
    understanding   47–48
stone(s)
    cabochon(s)   23, 24–25, 122, 127–145, 184
    chip(s)   24–25, 27, 46, 89, 92–103
    choices for   143–144
    setting   21, 122–125, 126–145
        irregular stone(s)   144

wire(s)
    filing the ends   63, 83
    marking for sawing   64
    preparing   78
    preparing to saw   63–64
    round   25, 27, 60, 62–63, 65–66, 69, 72, 76, 77, 78–79, 81–83, 87, 93, 94, 98, 119, 127, 132, 136, 156, 169, 172, 175
    square   25, 57, 61–66, 68, 69, 71–73, 77–78, 80–84, 86–87
    twisting   13, 76–87, 120, 135, 142, 184